Table of Contents

Introduction

    Preface 1

Student shorts

    Size of Infinite Sets 2

    Living Amid Probabilities 7

    Mathematical principles within famous painting 11

    Mathematical Biology: Introduction to Population Genetics 16

    Cryptography and Math 22

    Cycloids in Our Lives 28

    Pythagorean Theorems and Triples 34

    Introduction to Algorithms 38

    The Rising Importance of Mathematical Statistics in Economic Fields: The Stock Market 43

Preface

이 책에 대한 설명을 하려면 먼저 Math Circle 을 언급하지 않을 수 없습니다.

저는 2012 년 여름, 라스베가스에서 열린 ARML (American Regions Mathematics League)에 참관인 자격으로 초청을 받았습니다. ARML 에서 토론, 릴레이를 바탕으로 한 팀 수학 대회를 경험하였고, 학부모님과 대학 교수님들을 포함한 수많은 자원봉사자들과 다양한 국적의 학생들이 만드는 넘치는 생동감을 느꼈습니다. 그리고 이 둘의 공통 요소가 Math Circle 이라는 것을 알게 되었습니다. 그 당시 학생들을 평가해야만 하는 직업을 가진 성인 입장에서 만난 이 경험은 저에게 무한한 부러움과 무거운 책임감을 동시에 안겨주었습니다.

한국으로 돌아오는 비행기 안에서 미국과 다른 새로운 형태의 Math Circle 을 한국에서 만들어 보기로 마음먹었습니다.

첫째, 학생들이 스스로의 능력만으로 컨텐츠 생산자가 될 수 있게 돕는 것.

둘째, 학생들이 경쟁의 압박에서 자유로워지도록 돕는 것.

결론부터 언급하자면 이 결심으로 시작한 활동들의 결실인 이 책이 세상에 빛을 보기까지 6 년이라는 시간이 필요했습니다. 원고지에 쓴 글들이 바로 서적으로 출판할 수 없는 것과 같은 이유로 수년 전에 완성된 학생들의 자료들이 바로 책으로 나올 수 없었습니다. 공동 작업을 위한 템플릿 통합이 필요했고 실제 출판 과정들을 경험해 보지 않은 학생들의 이해가 필요했기 때문입니다.

이 책에 실린 자료들은 Mini Lecture, Student Seminars 등 다양한 기획의도를 통해 발전되었고 시행착오의 과정 끝에 경쟁을 배제한 Open Platform 구조를 완성할 수 있었다고 생각합니다.

운동을 시작한 첫날부터 올림픽에 참가하기를 희망하며 세계신기록을 세울 수 없듯이, 논문을 완성하고자 하는 학생들이 결과를 만들기 위한 과정의 소중함을 이해할 수 있도록 Open Research Forum 이 조력자로서의 역할을 잘 수행해 주기를 희망하고 응원하겠습니다.

2018 년 4 월
유충훈, Director of KGSEA

Size of Infinite Sets

Jae Hee Lee

Cheongshim International Academy

1. Introduction

In this mini-lecture, I would like to present how to study the size of infinite sets. What does it mean for an infinite set to have a size? How is it related to the usual notion of the size of finite sets? How do we compare the size of infinite sets? How big are the sizes of well-known infinite sets such as the natural numbers(N), the integers(Z), the rationals(Q), the reals(R)? There is a huge theory called the set theory to systematically study these topics, but in this lecture I will focus more on the interesting facts that will surprise you rather than rigorously going through all the details of set theory. Let us start with the discussion of finite sets, which has an intuitive notion of size which we are already familiar with.

2. Sizes of Infinite Sets

How do we know the size of finite sets? For finite sets, the definition of size captures the notion of the number of elements. If we denote the size of the set X as $|X|$, $|X|$ simply counts the number of elements in the set X. For example,

$$|\{1, 2, 3, 4\}| = 4$$

$$|\{1, 8, 56, 947, 2013\}| = 5$$

$$|\{dog, cat, elephant\}| = 3$$

Therefore, it is not difficult to think about the size of finite sets: you just count the number of elements in the set. It is also very easy to compare the size of two different sets. If two sets X and Y satisfy $|X| = m$ and $|Y| = n$ with $m < n$, then we could simply say that Y has more elements than X. Such an argument is quite valid and pretty much enough when we are dealing with finite sets. However, when we have to discuss infinite sets, the same approach works no more. What does it means to count the number of infinite sets? $N = \{1, 2, 3, \cdots\}$ obviously has no natural number corresponding to its size. Can we just say $|N| = \infty$? But we did not even define ∞. Even if we did, a new problem arises: how will we compare the sizes of infinite sets? Is $|N| = |\{2, 3, 4, \cdots\}| = \infty$? Is $|N| = |\{1, 4, 9, \cdots\}| = \infty$? Is $|N| = |\{2, 4, 6, \cdots\}| = \infty$? Do all infinite sets have the same size, namely ∞? To answer all these questions and to avoid contradictions, we have to examine the notion of the size of sets in a different way.

3. A New Way to Compare Sizes of Sets

Our new approach is to look at the functions from one set to another. Let X and Y be our sets, and see how functions can tell us about their relative sizes. A function is a mapping from the elements of X to the elements of Y. That is, for each x 2 X, wend its unique match, partner, or value from Y. We call this match f(x), and it is one of the elements 2Y. Saying that function f is from X into Y means that every element in X has a match in Y. Note that one element in X cannot have two matches, but the matches could be equal for different elements of X. Not all elements of Y have to be a match of some element of X. But if the two properties above actually do hold, then a function is special, and it reveals facts about the sizes of sets X and Y. Let us use our examples above. Let A = {1, 2, 3, 4}, B = {1, 8, 56, 947, 2013}, C = {dog, cat, elephant}. Note that A, B, C are sets, not the sizes of the sets. Let's define a function f from A to B, or more concisely f : A $\rightarrow$ B, as follows.

$$f(1) = 56$$
$$f(2) = 1$$
$$f(3) = 1$$
$$f(4) = 2013$$

This is really a function because every element in A has found its match in B. The matches are not equal, but that is okay. But this function is not really interesting for our discussion of the size of sets. So let us modify this function a little bit. Let us say f(3) = 8 instead of f(3) = 1. Then *different* elements of A has *different* matches in B. Such functions are called one-to-one. Because for one-to-one functions we have to find a *different* match for every element in A, we could see that B has to have at least as many elements in A, or we will run out of different elements in B to be a match of elements in A.

This is true for any sets! So if there exists a one-to-one function from set X to Y, $|X| \leq |Y|$.

This time let us look at the other property. Let us define a function g from A to C (g : A $\rightarrow$ C).

$$g(1) = cat$$
$$g(2) = dog$$
$$g(3) = cat$$
$$g(4) = elephant$$

This is certainly a function. Also, for every element in C, there is an element in A that has its match as that element. 1 and 3 has match cat, 2 has match dog, 4 has match elephant. That covers all of the elements in

C. A function satisfying such a property is called onto. Because for onto functions we have to find a match for every element in C, A should have at least as many elements in C, or we will run out of elements in A and some elements in C will not be matches of any elements in A. Remember that one element in A cannot have two matches, so for each element in C we need to find a different element in A.

This is also true for any sets! So if there exists a onto function from set X to Y, $|X| \geq |Y|$.

Or we could just say this is our definition of $\leq$ and $\geq$ for size of sets. From now on, we do not count the elements in the sets we are interested in, but we just see if we could find a special function from one to another. Basically, what we are doing here is choosing one element from each set and match them, until one of the set runs out of elements.

Especially, if we could find a function that is both one-to-one and onto, this would imply $|X| \leq |Y|$ and $|X| \geq |Y|$. In other words, $|X| = |Y|$. Now we have built a new way to compare the sizes of sets, and we could expand this definition to all sets. We could now compare the size of infinite sets!

4. Comparing the Size of Sets

Now that we have the appropriate machinery to work on sizes of any sets, let's get to work! First of all, we could now precisely define finite sets as sets with size equal to $|\{1, 2, \cdots, n\}|$ for some natural number n $\in$ N. We could also easily check that any finite set has smaller size than N, because we can never find an onto function from a finite set to an infinite set. We could also answer some of our original questions.

- Is $|N| = |\{2, 3, 4, \cdots\}|$? Yes, f(n) = n + 1 is both one-to-one and onto.
- Is $|N| = |\{1, 4, 9, \cdots\}|$? Yes, f(n) = n2 is both one-to-one and onto.
- Is $|N| = |\{2, 4, 6, \cdots\}|$? Yes, f(n) = 2n is both one-to-one and onto.

What? This is ridiculous! What we have just shown is that N has the same size as its subsets that lacks some elements in the original set of natural numbers(proper subsets). The whole has the same size as its small part. This may be quite counterintuitive, and even great mathematicians such as Gottfried Leibniz, the inventor of calculus, saw this fact as an outright contradiction. But using our new comparison standard, this turns out to be true. Infinite sets behave strangely.

In fact, even more unbelievable facts are waiting. We showed that some simple infinite subsets of N have the same size as N. It is not hard to show that any subset of N is either finite or has the same size as N. How about sets containing N?

In particular, let us look at the set of integers, Z = $\{\cdots, -2, -1, 0, 1, 2, \cdots\}$. If we could find a one-to-one, onto function from N to Z, then $|N| = |Z|$. This set seems to have at least twice the elements of N and the chance of finding such function seems miserable. However, if we define f : N $\to$ Z as

$$f(1) = 0$$

$$f(2) = 1$$

$$f(3) = -1$$

$$f(4) = 2$$

$$f(5) = -2$$

...

We could easily see that f is both one-to-one and onto. So |N| = |Z|.

But even more crazy results are waiting.

How about Q = { p/q : p ∈ Z, q ∈ N} Defining f : N → Q that is one-to-one and onto is more tricky. This time we order the rational numbers so that numbers using only 1 and −1 appears first, then the ones with 2 and −2, and so on. Of course we have to leave out the ones that appeared earlier. In other words, order the rationals in the following way:

$$\frac{0}{1}, \frac{1}{1}, \frac{-1}{1}, \frac{1}{2}, \frac{-1}{2}, \frac{2}{1}, \frac{-2}{1}, \frac{1}{3}, \frac{2}{3}, \frac{-1}{3}, \frac{-2}{3}, \frac{3}{1}, \frac{3}{2}, \frac{-3}{1}, \frac{-3}{2}, \frac{1}{4}, \frac{-3}{4}, \frac{4}{1}, \frac{4}{3}, \frac{-4}{1}, \frac{-4}{3} \ldots$$

This ordering is itself a function from N to Q. It is certainly one-to-one. It is also not hard to see that all rational numbers will some time appear in this list, meaning that the function is onto. So |N| = |Q|.

If you think about it for a while, this is a very surprising fact. Clearly N ⊂ Z ⊂ Q, but |N| = |Z| = |Q|. This can never happen in finite sets, but with our new definition of comparing the size of sets allows us to derive such interesting results.

At this point, you may be curious if there are any infinite sets strictly greater than N in size. Sets as large as the Q that is large enough to densely occupy the whole real line has the same size with N. What can come next? Our following discussion is about sets larger than the set of natural numbers.

5. |R| > |N|

The real numbers is the set that has a strictly larger size than N. How do we show this? There is an obvious one-to-one function from N to R, namely f(n) = n.

So certainly |N| ≤ |R|. If we cannot find a function that is both one-to-one and onto, then this will give us the fact that the reals are larger than the naturals.

In fact, we can never find an onto function from N to R. So particularly there are no functions that are both one-to-one and onto. Let us assume that there is an onto function f : N → R. If we could derive a contradiction from here, this will imply that our assumption was wrong, and therefore there are no such onto functions.

Since f is onto, each natural number has its real number match. This real number can be written in a decimal expansion: something like $3187.128398\cdots$, 1.25765, $0.00012382\cdots$. Now we construct a new real number as follows:

- Let this real number be between 0 and 1. So it is in the form of $0.?????\cdots$.
- Look at f(1)'s tenth(0.1) digit. If it is equal to 0, let the tenth digit of our number be 1. If it is not, let the hundredth digit be 0..
- Look at f(2)'s hundredth(0.01) digit. If it is 0, our number gets 1 for the
- hundredth digit. If it is not, our number gets 0.
- Look at f(3)'s thousandth(0.001) digit. If it is 0, our number gets 1 for the
- thousandth digit. If it is not, our number gets 0.
- Do the same thing for every n.

If we do this process, we get a new real number r with digits being either 0 or 1. Since f is onto, r must have a natural number n such that f(n) = r. However, this is impossible, because r and f(n) differs in the nth digit below 0. If f(n) has 0 at that digit, r has 1. If f(n) has any other number at that digit, r has 0. So this is a contradiction, and we cannot find an onto function from N to R. So $|N| < |R|$. This elegant argument is called the Cantor's Diagonalization.

So we have an infinite set larger than the set of natural numbers! In fact, we can find a set larger than $|R|$, a set larger than that set, and so on. There are many more interesting facts in set theory that are very surprising, but for this mini lecture I think this will be sufficient.

6. Miscellaneous (Technical Issues)

- Size of sets are usually called cardinality.
- Usually set theorists include 0 in N, but we did not.
- One-to-one functions are also called injections, and onto functions are also called surjections. If a function is both injective and surjective, it is called a bijection.
- Actually we were quite wrongfully using the notation $\leq$ and $\geq$ when we were talking about one-to-one functions and onto functions. It is much better to just say $|X| \leq |Y|$ if there is an one-to-one function from X to Y and define $\geq$ similarly. This is because of the Cantor-Bernstein-Schroeder Theorem which states that if there is an injection from X to Y and an injection from Y to X then $|X| = |Y|$. So using this definition we have $|X| \leq |Y|$ and $|X| \geq |Y|$ implies $|X| = |Y|$. This is not true for our definition unless we assume the Axiom of Choice, an extremely strong mathematical axiom.

Living Amid Probabilities

Joon-hyuk Chang, Minseok Kim
Korean Minjok Leadership Academy

1. Introduction

Greetings to all. It is with the utmost pleasure that we address you. Today, we will deliver a lecture on probabilities in our everyday lives. We have one small hope. We wish that this lecture will be memorable to most, if not all, of you, a unique lecture that many of you find striking. We will address topics such as lotteries, dice, and card games one at a time and undertake an examination at the probabilities involved in each of these issues. Sit back, relax, and enjoy our lecture. Let's begin.

2. Lotteries – Part 1

How much Lotto do you think you have to purchase before you win the first prize? Can you try to make a conjecture?

In case some of you are not familiar with how lotteries work, I would like to explain. When you buy a lottery, there are six numbers imprinted on it. The six winning numbers, which are announced periodically, are chosen randomly one at a time. How many same numbers you have determines the prize you will receive. If all the winning numbers are imprinted on the lottery you bought, congratulations: you have won the lottery.

Now, I would like to introduce you to my self-made lottery program.

Let's define a trial as the process of buying a lottery and checking the winning numbers. When I enter an integer between 1 and 6, inclusive, it runs a random simulation of consuming lotteries and counts the number of trials before there are as many matches between the winning numbers and the purchased lottery's numbers as the number I put as input. For example, if I enter 1, it shows me how many trials of buying lotteries I would have to undergo until one out of the six numbers match. The same goes with 2, 3, 4, 5, and 6.

How much time do you think will be needed for the program to indicate that I won first prize in the lottery? Let's begin running the program now with 6 as the input and continue discussing lottery more in depth near the end of the lecture.

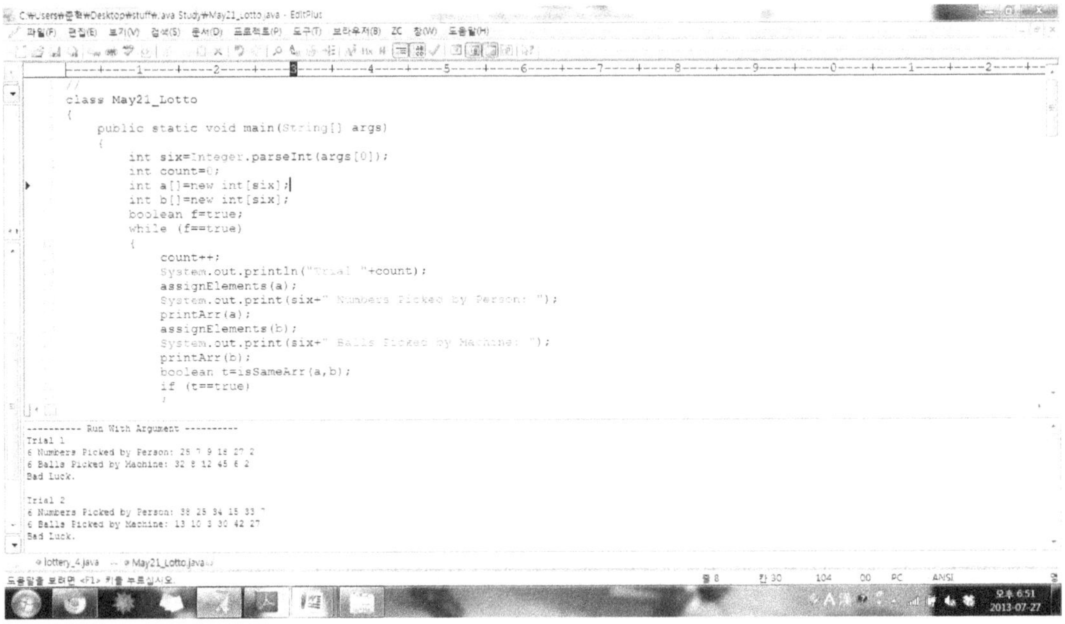

Figure 1: A lottery computer program

3. Dice

The six-sided dice has many properties that make it intimate with concepts of probability. Each of the six sides has distinct whole numbers from one to six inscribed on it. The theoretical probability of attaining each number when throwing a regular dice is 1/6.

*Theoretical Probability: the probability we would expect to observe

*Experimental (Empirical) Probability: the probability actually observed during testing processes

Understanding the distinction between theoretical probability and experimental probability is important. Throwing a large number of dice or throwing one dice multiple times illustrates the difference clearly. If, for instance, 12 dice are thrown, we would expect to get each of the six numbers two times. That is based upon theoretical probability. However, when we actually throw 12 dice, we can easily see that it is actually difficult to get each number exactly two times. Most likely, we are going to get some numbers more than others. The uneven distribution of the attained numbers can be easily noticed if we create six stacks of dice with each stack containing dice with the same number. After throwing the dice, if we calculate the probability that each number was attained, we will end up with the experimental probability.

An interesting probability I would like to point out is that the more dice rolled, the closer the experimental probability will be to the theoretical probability. For example, when we roll 36 dice and stack them into groups of same numbers again, we can notice that the lines are much more even then before. We receive a result much closer to what we can expect before actually throwing the dice.

4. Card Games

Cards are also items closely related to concepts of probability.

A typical deck of cards contain 52 cards, not counting the usually included two Jokers. In a deck, there are four suits – clover, spade, diamond, and heart – and thirteen ranks – numbers one through ten, Jack, Queen, and King. All of the cards are distinct, which means that although two different cards from the same deck may have the same suit or the same rank, they may not have the same suit and the same rank simultaneously.

Although all card games essentially and inadvertently involve probability to some degree, the two to discuss in this lecture are Poker and Blackjack.

a) Poker

Poker is a common gambling game involving betting. The person who can create the best combination of cards using his own cards and the cards on the table at the end of the game is declared the winner of a round of Poker.

When the game begins, the dealer gives two cards to each of the players. The cards that a player holds are called his 'hand.' Afterwards, three cards on the top of the deck are flipped so that every player can see them. The dealer flips another card, and the players take turns betting. This process is repeated until there is a certain amount of cards on the table. The players who had not folded reveal their hands, and the winner takes the pot.

Table 1: Poker probabilities

Hand	Frequency	Probability	Odds
Royal Flush	4	0.000154%	649,739:1
Straight Flush	36	0.00139%	72,192.33:1
Four of a Kind	624	0.0240%	4,165:1
Full House	3,744	0.144%	694.17:1
Flush	5,108	0.197%	508.80:1
Straight	10,200	0.392%	255.8:1
Three of a Kind	54,912	2.11%	47.33:1
Two Pair	123,552	4.75%	21.04:1
One Pair	1,098,240	42.3%	2.37:1
High Card	1,302,540	50.1%	1.96:1

The aspect of poker that makes it especially intimate with probability is the ranking order of the hands. The hierarchy is arranged in the order of the least likely outcome to the outcome with the highest probability.

The aspect of poker that makes it especially intimate with probability is the ranking order of the hands.

b) Blackjack

Blackjack is another well-known card game with an exceptional proximity to probability concepts.

The goal of Blackjack is to reach near as possible to 21. Each of the cards has an assigned value. Cards with ranks two to ten each has a value same as its rank. Jack, Queen, and King have values of ten. Ace can represent a value of either one or eleven, depending on how the player wants to use it.

The dealer deals each player two cards – one face-up, one face-down. Each player take turns deciding whether to receive more cards (also known as 'hit') or stop. The moment a player's hand exceeds 21, he is 'bust,' meaning that he is automatically out of the game. When all players have stopped, they reveal their hands, and the winner is the person whose hand's value is the closest to 21.

Blackjack is a game where the odds of any particular outcome can be calculated extremely accurately. By looking at the face-up cards of other players and your own cards, you can guess what cards are remaining in the deck. Thus, you can determine the approximate probability that you will not be bust when you hit. If there is a fair chance that you will get closer to 21 without exceeding it, it would be rational to hit. One must remember that there is a greater probability of being dealt a card of value 10 than cards of any other values.

It is interesting to note that the probability of being dealt a 21 – a Blackjack – is 4.83%. In a deck of 52 cards, there are four Aces and 16 cards with values of ten. Since you need one Ace and one card of value ten to get 21, there are 4*6=24 possibilities. The total number of outcomes can be calculated as corresponds to $_{52}C_2$ = 1326. Thus, the probability of being dealt a Blackjack is 64/1326 = 0.0483 = 4.83%. This particular percentage means that you will be dealt a Blackjack in about once every 21 rounds.

3. Lottery – Part 2

The probabilities of having a particular number of matches are as below (Table 2).

Thus, in order to win the first prize in the lottery, you would have to beat the one-in-8145060 probability.

You can notice that the Java lottery program is still running. At the rate it is operating, it will take about one day and a half before it wins the lottery.

Table 2: Lottery probability calculations

Number of Matches	Probability	Calculations
6	1/8145060	$\frac{1}{_{45}C_6} = 1 / \left(\frac{45*44*43*42*41*40}{6*5*4*3*2*1} \right)$
5	1/34807.9487	$\frac{_6C_5 * _{39}C_1}{_{45}C_6}$
4	1/732.7989	$\frac{_6C_4 * _{39}C_2}{_{45}C_6}$

Mathematical principles within famous painting

Dahee Chung

Cheongshim International Academy

1. Introduction

The goal of my lecture is to appreciate masterpieces and understand mathematical principles within each of pictures. We are going to skim masterpieces of various painters and catch mathematical tools and secrets from the pictures. After the brief explanation about famous paintings, we are going to understand mathematical meaning within the masterpiece and solve related problems. Aren't you interested? Let's start!

2. What is the golden ratio found in the body of 'The birth of Venus' and 'Hector and Andromache'?

The golden ratio (symbol is the Greek letter phi) is a special number approximately equal to 1.618. It appears many times in geometry, art, architecture and other areas. If you divide a line into two parts so that: the longer part divided by the smaller part is also equal to the whole length divided by the longer part. Then you will have the golden ratio.

Figure 1: *The birth of Venus* (left) and *Hector and Andromache* (right)

3. Related concepts

a) How many golden rectangles are in 'Broadway Boogie-Woogie'?

This rectangle has been made using the Golden Ratio, Looks like a typical frame for a painting, doesn't it?

Some artists and architects believe the Golden Ratio makes the most pleasing and beautiful shape.

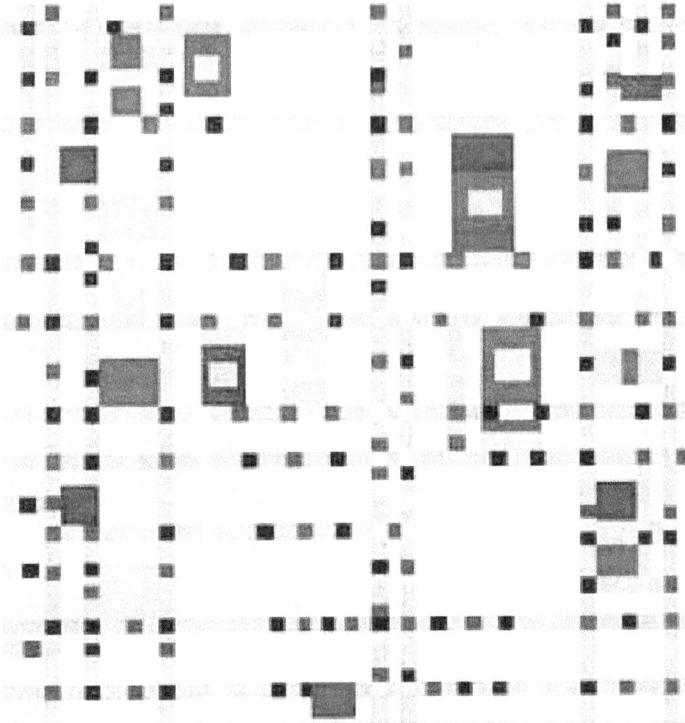

Figure 2: *Broadway Boogie-Woogie*

Here is one way to draw a rectangle with the Golden Ratio:

- Draw a square (of size "1")
- Place a dot half way along one side
- Draw a line from that point to an opposite corner (it will be $\frac{\sqrt{5}}{2}$ in length)
- Turn that line so that it runs along the square's side

Then you can extend the square to be a rectangle with the Golden Ratio.

b) A special relationship between the Golden Ratio and the Fibonacci Sequence

0, 1, 1, 2, 3, 5, 8, 13, 21, 34, ... The next number is found by adding up the two numbers before it. And here is a surprise: if you take any two successive *(one after the other)* Fibonacci Numbers, **their ratio is very close to the Golden Ratio.**

In fact, the bigger the pair of Fibonacci Numbers, the closer the approximation. Let us try a few:

A	B	B/A
2	3	1.5
3	5	1.666666666...
5	8	1.6
8	13	1.625
...	...	...
144	233	1.618055556...
233	377	1.618025751...
...	...	...

4. Can we draw the spirals in 'The shell' and 'Tower of Babel' with mathematical methods?

Figure 3: *The Shell* (left) and *Tower of Babel* (right)

In mathematics, a **spiral** is a curve which emanates from a central point, getting progressively farther away as it revolves around the point.

Two major definitions of "spiral" in a respected American dictionary are

a. A curve on a plane that winds around a fixed center point at a continuously increasing or decreasing distance from the point

b. A three-dimensional curve that turns around an axis at a constant or continuously varying distance while moving parallel to the axis; a helix.

We can build a square-ish sort of nautilus by starting with a square of size 1 and successively building on new rooms whose sizes correspond to the Fibonacci sequence:

Running through the centers of the squares in order with a smooth curve we obtain the nautilus spiral, the sunflower spiral. This is a special spiral, a self-similar curve which keeps its shape at all scales (if you imagine it spiraling out forever).

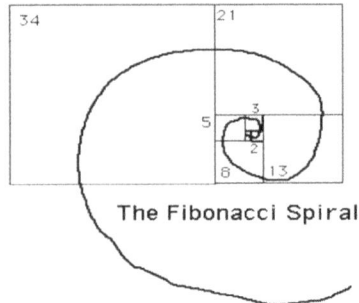

Figure 4: The Fibonacci Spiral

It is called equiangular because a radial line from the center makes always the same angle to the curve. This curve was known to Archimedes of ancient Greece, the greatest geometer of ancient times, and maybe of all time.

5. What is the secret of magic square hidden in 'Melencolia I' and 'Ssireum'?

A **magic square** is a square array of numbers consisting of the distinct positive integers 1, 2, ..., n^2 arranged such that the sum of the n numbers in any horizontal, vertical, or *main* diagonal line is always the same number, known as magic constant.

The 4 × 4 magic square, with the two middle cells of the bottom row gives the date of the engraving: 1514. This 4x4 magic square, as well as having traditional magic square rules, its four quadrants, corners and centers equal the same number, 34, which happens to belong to the Fibonacci sequence. His age in 1514 was 43, reverse of 34.

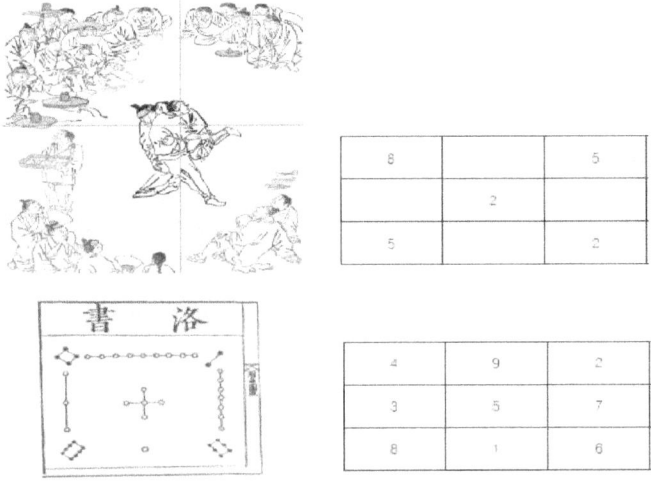

Figure 5: Magic squares in Ssireum (top) and *Melencolia I* (bottom)

6. Further Discussions

-What is the probability that swindler will win in *'The cheat with ace of diamond'*?

-What is the hidden meaning of pentagon and star in *'The Descent from the Cross'*?

-Where is the center of gravity in *'Woman holding a balance'*?

-What are secrets of *'The school of Athens'* and *'Brena Madonna'* which are drawn with the utilization of perspective?

-What symbolize the cabbalistic number and sign in *'Last Supper'*?

-What is volume relationship between the figures in *'Senecio'*

-Why the impossible triangle in *'Waterfall'* seems possible?

-What is the principle of melodious note in *'Spanish singer'* and *'Three musicians'*?

-*'The red son gnaws at the spider'* and dancing pi?

-What is the balance between circles within the *'Several Circles'*?

-What is the meaning of clock in *'The Persistence of Memory'*?

-Why the legs of easel in *'The Human Condition'* and *'atelier'* are three?

Mathematical Biology: Introduction to Population Genetics

Philjae Chang

Cheongshim International Academy

1. Basic Principles and Vocabulary in Genetics

<u>Genes and Alleles</u>

Genes are molecular units of heredity of a living organism. Allele is one of alternative forms of the same gene or same genetic locus.

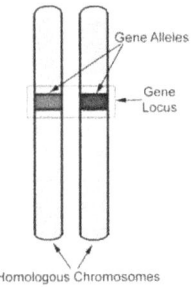

Figure 1: Genes and Alleles

<u>Genotype and Phenotype</u>

Genotype: outward, physical manifestation; composes of two alleles

Phenotype: internally coded, inheritable information

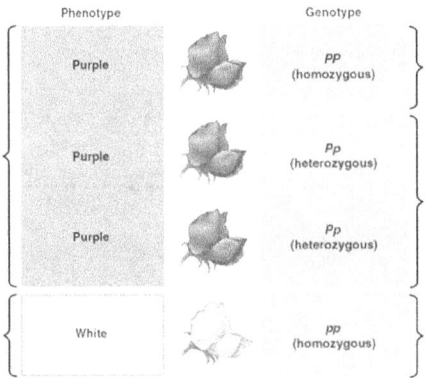

Figure 2: Genotype and Phenotype

Dominant and Recessive Alleles

Dominance in alleles determines the expression of traits. Each allele could be either dominant or recessive. If a dominant allele exists in a gene, the phenotype follows the dominant form. If there isn't any dominant allele but only recessive alleles, the phenotype shows the recessive form.

Punnet Square

Punnet Square is a simple grid with all possible male gamete genes and female gamete genes. The grid can be completed by filling in square with the genotype that can be generated from each combination of the row and the column.

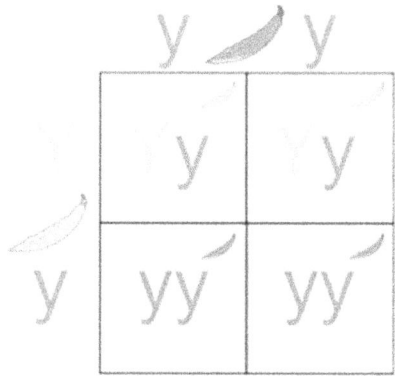

Figure 3: Punnet Square

2. Basic Math in Genetics

Possibilities of Hereditary: Hardy – Weinberg Principle

Requirements:

a) No selection

b) No mutation

c) Infinitely large population

d) No emigration or immigration

e) Random mating

The Hardy – Weinberg Equilibrium happens when

p= frequency of 'A allele'

q= frequency of 'a allele'

and
$$p + q = 1$$
$$(p + q)^2 = 1$$
$$p^2 + 2pq + q^2 = 1$$

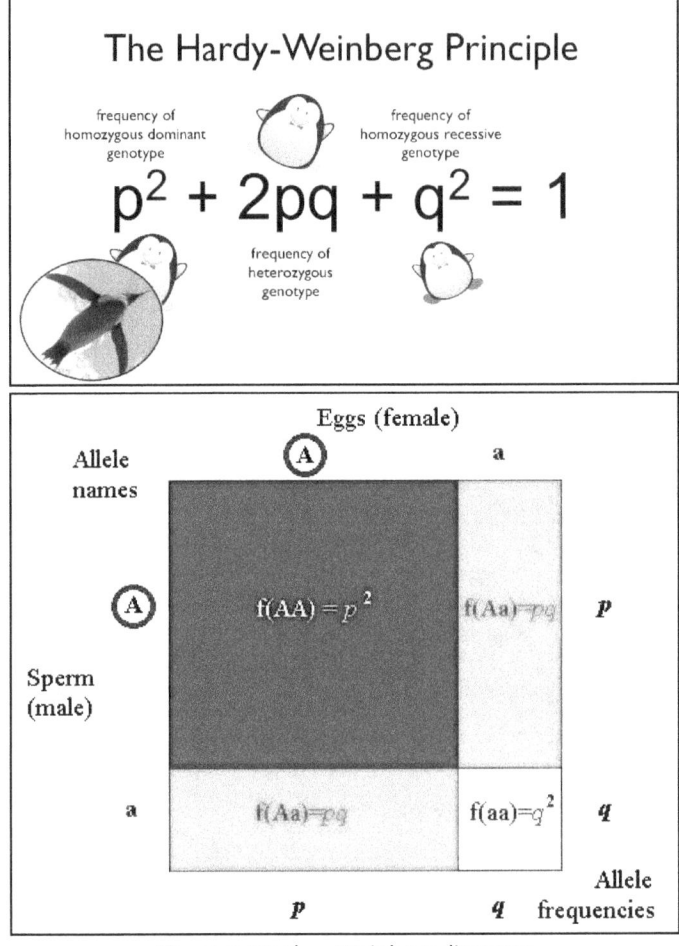

Figure 4: Hardy – Weinberg diagrams

3. Effects of Selection

p_t = Relative frequency of **A** allele in current generation ($0 \leq p_t \leq 1$)

q_t = Relatice frequency of **a** allele in current generation ($0 \leq q_t \leq 1$)

w_{AA} = Relative survival of AA genotype

w_{Aa} = Relative survival of Aa genotype

w_{aa} = Relative survival of aa genotype

N= The real population size

Genotype frequency= Different from a frequency of an allele, it means the frequency of genotype itself, such as Aa or AA.

According to Hardy – Weinberg Equilibrium, in current generation t,

$$q_t + p_t = 1$$

$$q_t^2 + 2p_t q_t + p_t^2 = 1$$

After the time goes on, in the next generation t+1, the relative frequencies will change, so let's multiply each genotype frequency by its relative survival.

$$AA: p_t^2 w_{AA}$$

$$Aa: 2p_t q_t w_{Aa}$$

$$aa: p_t^2 w_{aa}$$

(1)

The relative frequencies will have relationships below.

$$p_{t+1} + q_{t+1} = 1$$

Average survival $\overline{w_t}$ in t+1 will be

$$\overline{w_t} = p_t^2 w_{AA} + 2p_t q_t + q_t^2 w_{aa}$$

Therefore, the new relative genotype frequencies, whose sum is out of 1, will be the frequencies in (1) divided by the average survival $\overline{w_t}$.

$$AA: p_t^2 \frac{w_{AA}}{\overline{w_t}}$$

$$Aa: 2p_t q_t \frac{w_{Aa}}{\overline{w_t}}$$

$$aa: q_t^2 \frac{w_{aa}}{\overline{w_t}}$$

(2)

Meanwhile, the frequencies in the next generation t+1 can be determined, using similar principle in Punnet Square.

$$p_{t+1} = \frac{2N \cdot frequency\ of\ AA + N \cdot frequncy\ of\ Aa}{2N}$$

$$= frequency\ of\ AA + \frac{1}{2}(frequency\ of\ Aa)$$

$$= \frac{p_t^2\ w_{AA} + \frac{1}{2} \cdot 2p_t q_t\ w_{Aa}}{\overline{w_t}} = \frac{p_t(p_t\ w_{AA} + q_t w_{Aa})}{\overline{w_t}}$$

Define Δ_p as the change in p from generation to $t+1$,

$$\Delta_p = p_{t+1} - p_t = \frac{p_t(p_t w_{AA} + q_t w_{Aa})}{\overline{w_t}} - p_t$$

$$= \frac{p_t}{\overline{w}}(p_t w_{AA} + q_t w_{Aa} - \overline{w_t}) = \frac{p_t}{\overline{w}}[p_t w_{AA} + q_t w_{Aa} - (p_t^2 w_{AA} + 2p_t q_t + q_t^2 w_{aa})]$$

$$= \frac{p_t}{\overline{w_t}}[w_{AA} \cdot p_t(1-p_t) + w_{Aa} \cdot q_t(1-2p_t) - q_t^2 \cdot w_{aa}]$$

$$= \frac{p_t}{\overline{w_t}}[w_{AA} \cdot p_t q_t + w_{Aa} q_t(1-p_t-p_t) - q_t^2 \cdot w_{aa}]$$

$$= \frac{p_t}{\overline{w_t}}[w_{AA} \cdot p_t q_t + q_t(q_t - p_t) \cdot w_{Aa} - q\_t^2 \cdot w_{Aa}]$$

$$= \frac{p_t q_t}{\overline{w_t}}[w_{AA} \cdot p_t + (q_t - p_t) \cdot w_{Aa} - q_t \cdot w_{Aa}]$$

$$= \frac{p_t q_t}{\overline{w_t}}[p_t \cdot (w_{AA} - w_{Aa}) + q_t \cdot (w_{Aa} - w_{aa})]$$

(3)

From equation (2) above,

$$\overline{w} = p_t^2 w_{AA} + 2p_t q_t w_{Aa} + q_t^2 w_{aa}$$

To express q in terms of p,

$$\overline{w} = p_t^2 w_{AA} + 2p_t(1-p_t)w_{Aa} + (1-p_t)^2 w_{aa}$$
$$= p_t^2 w_{AA} + 2p_t w_{Aa} - 2p_t^2 w_{Aa} + w_{aa} - 2p_t w_{aa} + p_t^2 w_{aa}$$

Then take the derivative of this to get an expression for how the average fitness of the population changes as the allele frequencies change.

$$\frac{d}{dp}\overline{w} = 2p\, w_{AA} + 2w_{Aa} - 4pw_{Aa} + 0 - 2w_{aa} + 2p\, w_{aa}$$

$$= 2(p\, w_{AA} + w_{Aa} - 2p\, w_{Aa} - w_{aa} + pw_{aa})$$

$$= 2[p(w_{AA} - W_{Aa}) + (1-p)(w_{Aa} - w_{aa})]$$

(4)

However, note that $\Delta_p = \frac{pq}{\overline{w_t}}[p_t(w_{AA} - w_{Aa}) + q_t(W_{Aa} - w_{aa})]$

Then,

$$\Delta_p = \frac{p_t q_t}{2\overline{w_t}} \cdot \frac{d}{dp}\overline{w_t}$$

(5)

According to (5),

While pq is proportional to the variance in allele frequencies,

$\frac{d}{dp}\overline{w_t}$ shows the strength of selection: how strongly fitness varies.

The rate of evolution is proportional to the product of

$$\text{Genetic variability} \times \text{strength of selection}$$

At the non-trivial equilibrium, with both p and q greater than 0, $\Delta_p = 0$ requires that $\frac{d}{dp}\overline{w} = 0$.

In conclusion, natural selection maximizes the mean fitness $\overline{w}$.

Cryptography and Math

Yongchan Hong

Cheongshim International Academy

1. Introduction

Hello ladies and gentlemen! Welcome to the world of cryptography! Cryptography is an essential field of study that is utilized in everyday life. Even though it is not shown directly, cryptography is used in WIFI, Bitcoin, SSL, digital signatures and etc. You cannot imagine a world without using cryptography.

Then what is exactly "Cryptography"? There are various definitions for cryptography, but I chose the most clear and concise one. Cryptography is a method of storing and transmitting data in particular form so only those for whom it is intended can read and process it. This cryptography had started long from ancient times, Rosetta stone used in order to keep its information secret. In order to keep each person's information safe, cryptography uses mathematics in various ways. In this lecture, we will cover mathematics applied in basic cryptography, such as Caesar, Affine Caesar, Hill, Rijndal and RSA.

2. Classification of Cryptography

Cryptography is a subject that has various fields inward. Cryptography is mainly divided into two, which is transposition and substitution. Transposition is changing the word order of the message. Substitution is changing the word itself. This substitution is divided into two, which is by code and cipher. Code is changing the word itself, while cipher is changing single characters. In this lecture, we will only discuss about coding in cipher.

Cryptography	Transposition	
	Substitution	Code
		Cipher

3. Basics for Cryptography

Before we learn how math is used in cryptography, we should learn about basic function of cryptography. Cryptography uses P, C, E and D to express its function. P is plain text. Plain text is the text that we wish to code. C is cipher text. Cipher text is the encoded plain text. E is encryption function, which is the function that we use in order to change plain text to cipher text. D is decryption text, which is the function that we use in order to change cipher text to plain text.

We show this function as:

$$E(P)=C$$

$$D(C)=P$$

To give example, let us say plain text is "APPLE." If we move each alphabet to next alphabet, we can encode this into "BQQMF." This text is cipher text. So,

$$E(APPLE)=BQQMF$$

$$D(BQQMF)=APPLE$$

As we now understand the basics of cryptography, let us dive into mathematics that resides inside the cryptography.

4. Caesar Cipher

The Caesar cipher is one of the early known and simplest encoding. It was used by Julius Caesar in protecting military messages. By using Caesar Cipher, Caesar could be successful in Rome. Caesar cipher is shifting the number by certain amount "t". It is usually used in alphabet, and it is encoded by:

$$A\to 0,\ B\to 1,\ C\to 2,\ D\to 3\ldots Z\to 25$$

And shift each number into certain amount and get new alphabet. The case "APPLE" I used to explain in "Basics for Cryptography" is also using Caesar Cipher. To explain in function,

$$E(P)=(P+t,\ \mathrm{mod}\ M)$$

$$D(C)=(C-t,\ \mathrm{mod}\ M)$$

Let me explain what this means by using the "APPLE" example again. M in this place is 26, as we are using alphabetical order and changing those numbers into 0–25. If we say t=2, we should move our alphabets into 2 next alphabet.

$$E(P)=(P+2, \bmod 26)$$

$$\text{As } A=0, E(0)=(2, \bmod 26)=C$$

$$\text{As } P=16, E(16)=(18, \bmod 26)=R$$

$$\text{As } L=12, E(12)=(14, \bmod 26)=N$$

$$\text{As } E=5, E(5)=(7, \bmod 26)=G$$

$$E(APPLE)=CRRNG$$

So, we can get Cipher Text of CRRNG. In order to decipher this Cipher text CRRNG, we can simple subtract 2 (which is t) and get out plain text back. Caesar Cipher is most simple encoding that we can do.

5. Affine Caesar Cipher

This cipher is going one step further than Caesar Cipher. We multiply certain number "a" in plain text and add "t". To show in function:

$$E(P)=(aP+t, \bmod M)$$

$$D(C)=((C-t)/a, \bmod M)$$

Caesar Cipher is the special case when a is 1. Let us use "APPLE" again. This time, we can encode by a=2, and t=1.

$$E(P)=(2P+1, \bmod 26)$$

$$\text{As } A=0, E(0)=(1, \bmod 26)=B$$

$$\text{As } P=16, E(16)=(33, \bmod 26)=H$$

$$\text{As } L=12, E(12)=(25, \bmod 26)=Z$$

$$\text{As } E=5, E(5)=(11, \bmod 26)=K$$

$$E(APPLE)=BHHZK$$

This Affine Caesar Cipher is hard to decode than Caesar Cipher. Now, we will look through other ciphers that are complex.

6. Hill Cipher

Hill Cipher is introduced in 1929 by Lester Hill, and this Hill cipher is a poly-alphabetic cipher that uses matrices to encode plaintext messages. This cipher is using certain n x n key matrix that is pre chosen.

In this cipher, we will use 2 x 2 matrix $\begin{bmatrix} a & b \\ c & d \end{bmatrix}$ that is invertible mod 26. We should acknowledge that Det(A)=ad-bc (mod 26).

This time, as it should be multiple of 2, let us encode "LOVE" by using key matrix $\begin{bmatrix} 3 & 7 \\ 9 & 10 \end{bmatrix}$. If we change LOVE in number, it is 12, 15, 22, 5. If we perform Hill cipher:

$$\begin{bmatrix} 3 & 7 \\ 9 & 10 \end{bmatrix} \begin{bmatrix} 12 \\ 15 \end{bmatrix} = \begin{bmatrix} 141 \\ 258 \end{bmatrix} \text{ mod } 26$$

$$\begin{bmatrix} 3 & 7 \\ 9 & 10 \end{bmatrix} \begin{bmatrix} 22 \\ 5 \end{bmatrix} = \begin{bmatrix} 101 \\ 248 \end{bmatrix} \text{ mod } 26$$

So, the numerical value of the cipher text is 11, 24, 23, 14

We can get cipher text KXWN

This hill cipher looks complex compare to other encoding methods! Then how can we decode Hill cipher? Decrypting the Hill Cipher is quite complex first, we should use A^{-1} matrix. A^{-1} is $\det(A)^{-1} \begin{bmatrix} d & -b \\ -c & a \end{bmatrix}$ mod m. We should then multiply cipher text in order to get plain text again. If we perform decrypting action:

$$A^{-1} = 19^{-1} \begin{bmatrix} 10 & -7 \\ -9 & 3 \end{bmatrix} \text{ mod } 26$$

$19^{-1} = 11$, as $19*11 = 1$ mod 26

$$A^{-1} = \begin{bmatrix} 110 & -77 \\ -99 & 33 \end{bmatrix} = \begin{bmatrix} 6 & 1 \\ 5 & 7 \end{bmatrix} \text{ mod } 26$$

If we put cipher text 11, 24, 23, 14

$$\begin{bmatrix} 6 & 1 \\ 5 & 7 \end{bmatrix} \begin{bmatrix} 11 \\ 24 \end{bmatrix} = \begin{bmatrix} 90 \\ 223 \end{bmatrix} \text{ mod } 26$$

$$\begin{bmatrix} 6 & 1 \\ 5 & 7 \end{bmatrix} \begin{bmatrix} 23 \\ 14 \end{bmatrix} = \begin{bmatrix} 152 \\ 213 \end{bmatrix} \text{ mod } 26$$

We can get plain text number 12, 15, 22, 5

We can get plain text LOVE.

This Hill cipher could be used in any word or phrase that is multiple of n. By cutting in n words, we can encode in n x n matrix. How is this ciphering method? Now we will talk about the most complex ciphering method: RSA.

7. RSA Cipher

RSA cipher is the first encryption that made electric signal possible. This encryption was made by Ron Rivest, Adi Shamir and Leonard Adleman in 1977, and based on their name the "RSA" name has been made. Unlike other encrypting methods, RSA methods use two keys. This key means constant that can open and close the message. There is public key that is used to encrypt message and private key that is allowed to certain person to decrypt it. We should first understand about key generation.

1. We should choose p and q that is two distinct co-prime numbers.
2. Compute N=pq
3. Compute $\varphi(N)=(p-1)(q-1)$. This φ is Euler's totient function. This value is private.
4. Search e that is smaller than $\varphi(N)$ and co-prime to $\varphi(N)$.
5. Search d that is de≡1 mod $\varphi(N)$

In here, (N, e) is public key and (N, d) is private key. It is important to erase p and q in order to prevent someone to guess d and e.

For encryption, we undergo this procedure:

$$C=p^e \bmod N$$

For decrypting, we undergo this procedure:

$$P=c^e \bmod N$$

This C is cipher text, and P is plain text. Do not forget this! This is quite confusing? So we will now discuss about actual performance for RSA.

1. Select two different prime numbers.

   Select p=61, q=53

2. Compute N=pq

   N=61 x 53=3233

3. Compute $\varphi(N)=(p-1)(q-1)$.

$$\varphi(3233)= (61-1)(53-1)= 3120$$

4. Select any number e between $1<e<3120$ and e that is prime to $\varphi(N)$

$$\text{Select } e=17$$

5. Compute d by the formula that is shown above.

$$17 \times 2753= 46801= 1 \pmod{3120}$$

$$d= 2753$$

6. Encrypt and decrypt using d and e.

For example, plain text p= 65 can be encrypted as:

$$C=65^{17} \pmod{3233}= 2790$$

Cipher text C= 2790 can be decrypted as:

$$P=2790^{2753} \pmod{3233}= 65$$

This RSA cipher is quite complex, right? And now we learned one of the essential math that is used in Cryptography!

3. Conclusion

As you see, this cryptography is based on the mathematics. If mathematics wasn't used, such development of cryptography might not be possible. Yes, our whole world is filled with mathematics even though it is not appeared directly.

Cycloids in Our Lives

Hyunji Kim

Seoul International School

1. Introduction

There are many wonders in our world, both from the technological innovation of mankind and the technicalities of the natural world that we simply take for granted. But upon closer look, you will be able to find many math concepts that are hidden behind. Today, I would like to focus on one of these hidden concepts, which is very pervasive in our lives but only a few know about. Well, without further a due, let us begin!

2. What is a Cycloid?

First, let us start off with a simple question.

"What is the shortest distance from one point to another?"

No one will have trouble coming up with the answer. It's a straight line.

Let me ask you a slightly different question.

"What is the *fastest* way from one point to another?"

Now you may be thinking *'That's also a straight line!'*

But in fact, it is not. Think back at some of the slides you have ridden. Were they all straight? You may recall that the fun, or the more thrilling ones were those that were curved. This is because they were faster. As you may have guessed, the fastest way between two points is reached through a *curve*. Not just any random curve, but a special curve called a 'Cycloid'.

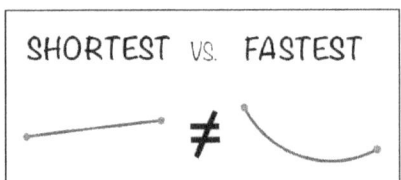

Figure 1: Shortest vs Fastest

So what exactly, is a cycloid?

By definition, it is a curve generated by a point on a circumference of a circle that rolls without slipping, on a straight line.

As you can sense from the definition, drawing a cycloid is very simple. All you need to do is follow what the definition tells you to do.

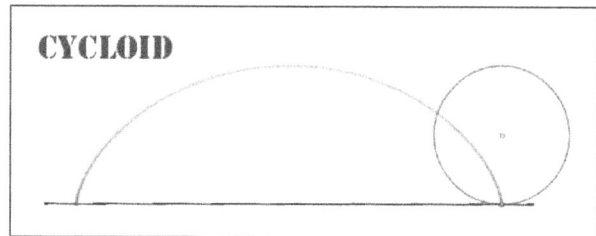

Figure 2: A cycloid

To create your own, take these two steps.

Mark the point where the circle and line meets.

Rotate the circle on the straight line. As you do so, mark the path that the point takes.

You can mark the path by fixing a pencil on the point so that it naturally creates a path as you rotate the circle.

You can also mark the points at different intervals while rotating the circle and connect the points at the end to make the curve.

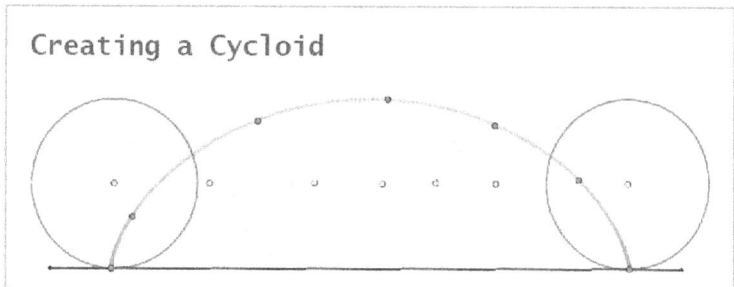

Figure 3: Drawing a cycloid

But this seemingly simple concept is not simple at all, earning the title, 'Helen of Geometry'.

Helen was a princess that caused the Troy War. Likewise, the cycloid causes frequent quarrels with even the founder of the cycloid still remaining a mystery.

With its many mysterious aspects, the curve's traits are unique and numerous as well.

3. Characteristics of a Cycloid

For now, let us focus on two main characteristics, that we will see being applied in our everyday lives.

These characteristics should be taken with a picture of an inverted cycloid. You will find that an upright cycloid will need energy for something on one point to even start moving.

The first is the obvious. As I have mentioned before, the cycloid is a way of connecting two points so that they take the shortest amount of time to travel. Hence, the cycloid has been given the name *'brachistochrone'*.

The second, and more unique characteristic gives the cycloid another name, called *'tautochrone'*. What this is that wherever one starts on the curve, it will always fall to the bottom of the curve at the same time.

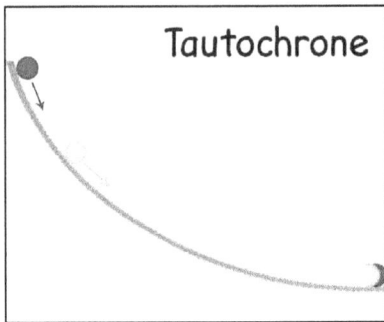

Figure 3: A tautochrone

4. Cycloid in our Daily Lives

How are these cycloids seen in our daily lives? We will take a look into how cycloids have become part of nature through evolution. Cycloids can also be seen in the human society, from the time we did not even know about the concept of cycloids to after we found out about the characteristics and put it to use.

a) A Closer Look into Nature

There are a great number of animals and plants on our planet. Today, however, it is difficult to see many in our daily lives.

But one that we always see, even in the cities, is the bird. They waken us up with their songs and sing at regular intervals. As you may have already guessed what I am trying to get at, these birds use cycloids as well, although it is instinctive.

When birds see their prey, whether it is bugs for sparrows or mice for eagles, they have to get to the prey as quick as possible, or else the prey might run away or be taken by another predator. Hence, **birds make a cycloid path from their position to the prey's position.** These do not apply to all birds, however. It is usually seen in birds such as eagles and owls.

Cycloids can also be seen in the anatomy of birds and fish.

They can also be seen in the **scales of fish**. Especially, they are usually seen in bony fishes. These scales allow the fish to move faster through the water as the cycloids allow the water to pass quickly. For example trout, which lives in the fast waters have cycloid scales.

Figure 4: Cycloids in nature

For us, these cycloids would take some amount of time to work out, as we would have to measure the distance, find the circle with the circumference matching the distance, and draw a cycloid. However, birds instinctively take cycloid paths instantly, without even having to think. Also, fishes are already born with these curves. This is able to take place through evolution, as animals try to maximize their ability to survive.

b) A Closer Look into Our World

In our technological world, we have made great use of cycloids. However, just like the instinctive natures of animals, our ancestors have used these concepts as well.

Let me first introduce you to **_Korean traditional roof tiles_**.

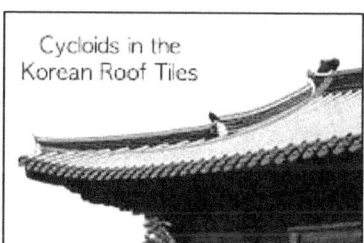

Figure 5: Cycloids in Korean Roof Tiles

Koreans made roof tiles that were curved in order to avoid decay. They had to find ways to make the roof not let in any water. They best way, they found would be to make the water run down the roof fast enough so that it would not seep in. While they could have made it straight, through many generations, they made the tiles curved, which resemble that of the cycloid and minimized the level of decay.

Figure 6: Sungnyemun

A great example of the cycloid roof's effect showed, sadly, at a devastating incident at Sungnyemun. This place was built in the 1390s, as one of the eight gates in the fortress wall of Korea, and the oldest standing monument in Seoul.

Figure 7: Fire in Sungnyemun

On February 10th, 2008, a small fire started at Sungneymun. Even though the fire was found at an early stage, it could not be put out for five hours because of the cycloid tiles repelling water. Even though the water was poured, it simply 'rubbed off' too fast, and hence could not go in.

Our ancestors made great use of roof tiles, to keep their houses from decaying. The effect of cycloids demonstrated too well, however, and ended up in a disaster.

(On a side note, the Sungnyemun was rebuilt.)

Moving on from how our ancestors used cycloids, let us take a look at how we are using cycloids today, in our modern world.

Figure 8: Cycloids in slides

As I mentioned briefly in the beginning, cycloids are closer to us then we think it to be. A great example is a *slide*.

There are many types of slides, but most have curves in them, which is more fun. This is because the curves are cycloids, which make us go down faster, letting us feel more thrill. The next time you go on a slide, see what kind it is. If there is a straight one and a curved one, try it out to feel which one is faster.

Another example can be seen in the *path of spaceships*.

In order to get into space using the least amount of fuel and hence the quickest amount of time, spaceship's paths are usually in the form of a cycloid.

Lastly, these cycloids are used in **gears in clocks.**

Gears, which are used in many machines, vary in type. Today, gears in most machines are involute gears since they are easier to make and cheaper. While they hold benefits, there are a lot of interruptions among these gears as the convex parts come in contact with the convex parts creating parts where they are not connected.

In the case of objects that require meticulousness such as clocks or cameras, there cannot be any interruptions.

Here, cycloid gears, which are harder to make are useful. To be exact, it is the cycloids called *epicycloids* (cycloids drawn not on a line but around the 'outside' of a circle), and *hypocycloids* (cycloids drawn around the 'inside' of the circle) that allows for no interruptions. These cycloids make up the flank (side of the teeth), with the convex epicycloids on top and concave hypocycloids on the bottom. Hence, unlike convex meeting convex, the epicycloid meets the hypocycloid, and fits perfectly continuously, which allows for no interruptions.

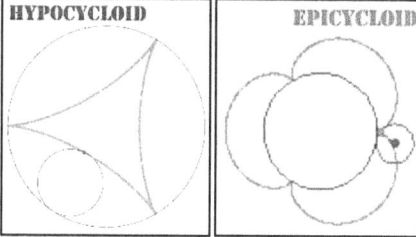

Figure 8: Hypocycloids and epicycloids

5. Conclusion

Cycloids are just one small part of the large world of mathematics. Rather than just seen in a math textbook however, cycloids are in our every day lives, helping and entertaining us. Cycloids are not the only math principles that are applied in our lives. With just a little amount of interest, you will find countless applications of different principles in math.

Now, take a *closer* look around. What do you see?

Pythagorean Theorems and Triples

Donghun Cho

Korean Minjok Leadership Academy

1. Introduction

Hello, everyone! The brief speech that I will give to you is about very well-known geometric theorem, Pythagorean Theorem. This very basic theorem, as you do know already, is very useful tool for solving many geometric problems. Today, with this interesting theorem, I would like to present to you some basic knowledge and proofs of this theorem, Pythagorean triples, and how this theorem can be used in all kinds of area. Also, more importantly, I want to show you that you can get so much intriguing facts out of very simple and basic theorem. Surprising power of math.

2. Proofs of Pythagorean Theorem

Before I show different types of proofs, I would explain what the Pythagorean Theorem is, although you would probably all know this already. Pythagorean Theorem states that the square of the hypotenuse (the side opposite to the right angle) is equal to the sum of the squares of the other two sides.

$$a^2 + b^2 = c^2$$

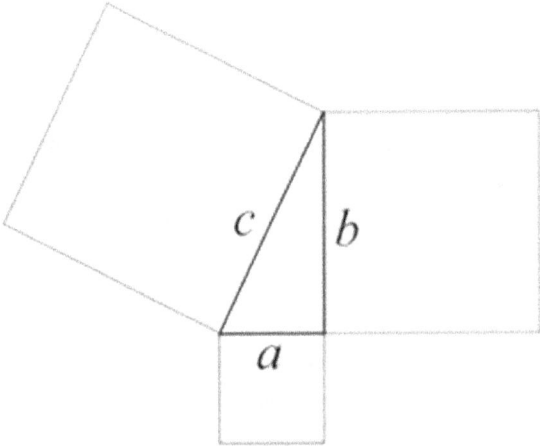

Figure 1: The Pythagorean theorem

Now, I will show you few different proofs of this fundamental thorem

a) Geometric Proof

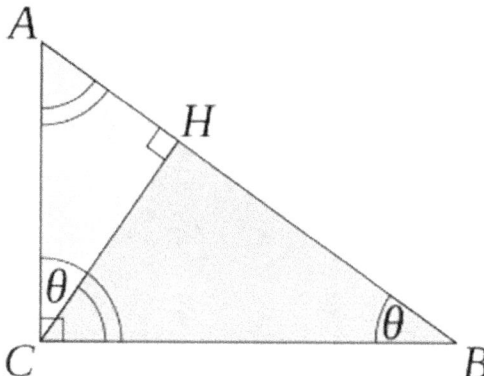

Figure 2: Geometric proof

From the figure above, you can see that ABC, ACH, and CBH are all similar triangles. So $\frac{BC}{AB} = \frac{BH}{BC}, \frac{AC}{AB} = \frac{AH}{AC}$. $BC^2 = AB \times BH$ and $AC^2 = AB \times AH$. By adding these two equations, we get $BC^2 + AC^2 = AB \times (BH + AH) = AB^2$. Finally, we can conclude $AB^2 = AC^2 + BC^2$.

b) Algebraic Proof

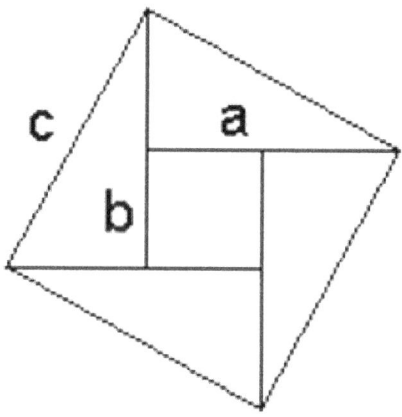

Figure 3: Algebraic proof

(Area of big square) = (Area of small square) + 4 × (Area of triangle)

$$c^2 = (a-b)^2 + 4 \times \frac{ab}{2}$$

$$c^2 = a^2 + b^2$$

c) Differential Proof

The two proofs I have presented above are very commonly used proofs, but the proof I will show right now is not introduced very often. This proof is using differential equation.

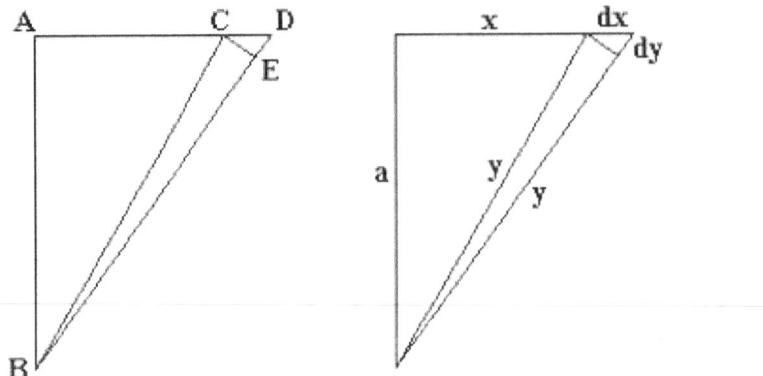

Figure 4: Differential proof

ABC and ECD are similar triangles.

$$\frac{x}{y} = \frac{dy}{dx}$$

$$y\,dy - x\,dx = 0$$

By solving differential equation(doing integration on both part), we get

$$y^2 - x^2 = C$$

The constant can be determined from the initial condition for x=0.

Since y(0)=a, $C = a^2$

So, $y^2 = x^2 + a^2$

Now, I have displayed only three proofs of the Pythagorean Theorem. All of the proof are approached from different ways, geometric, algebraic, and differential. However, it is surprising that there are still 40 more proofs that have been discovered. It is fascinating that there could be 43 proofs of one little thorem.

3. Pythagorean Triples

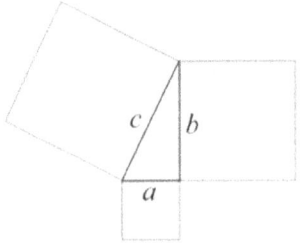

Figure 1 *Revisited*

In this section, I would like to talk about Pythagorean Triples. A Pythagorean triple consists of three positive integers a,b, and c, such that $a^2 + b^2 = c^2$. Such a triple is commonly written (a,b,c), and some of the examples are (3,4,5), (5,12,13), (8,15,17), and so on. I would like to show that these numbers are not just random, they have certain patterns.

All triples can be written as (a,b,c) where $a^2 + b^2 = c^2$, a,b, and c are coprime, and b and c have opposite parities(one is even and one is odd). If b and c are both even, then a,b, and c are not coprime. If b and c are both odd, the equation would equate an even to an odd. So b and c must have opposite parities. Since $a^2 + b^2 = c^2$, we can get $c^2 - a^2 = b^2$ and so $(c-a)(c+a) = b^2$. So, $\frac{a+c}{b} = \frac{b}{c-a}$. Since $\frac{a+c}{b}$ is rational, we can set it to equal $\frac{m}{n}$ in lowest terms, then $\frac{b}{c-a} = \frac{m}{n}$. So $\frac{c-a}{b} = \frac{n}{m}$. $\frac{c}{b} + \frac{a}{b} = \frac{m}{n}$ and $\frac{c}{b} - \frac{a}{b} = \frac{n}{m}$.

By adding and subtracting two equations above, we can get $\frac{c}{b} = \frac{m^2+n^2}{2mn}$ and $\frac{a}{b} = \frac{m^2-n^2}{2mn}$. Finally, we can see that a = $m^2 - n^2$, b = $2mn$, and c = $m^2 + n^2$ satisfy the Pythagorean Theorem. If you first look at the lists of the Pythagorean triples, you might think that you actually have to do the trial and error for numerous numbers and check if it works or not. However, there are much simpler way than calculating all of the numbers.

4. Conclusion

In this very brief presentation, I have dealt with quite fundamental theorem, Pythagorean Theorem. I wanted to show you that even with this commonly used theorem, if you think just a little bit deeper, you can learn more interesting things. Although it wasn't a very professional presentation, I hope you gained little more interest in mathematics than you had before.

An introduction to algorithms

Junghwan Lee
Cheongshim International Academy

1. Introduction

In order for a computer to do anything, it needs a computer program to run. To write a computer program, you need to tell the computer exactly what to do. When a computer program is written and inputted into the computer, the computer will then follow the rules and steps of the computer program exactly as you wrote it to accomplish something. Inside the computer programs are the algorithms. Algorithms are the set of well-defined procedures that allows the computer to accomplish a goal.

An algorithm should not ever be an ambiguous statement that varies when repeated infinitely or can be subjectively interpreted. If that is case, the program will have a bug.

For the sake of simplicity, many of the algorithms below use general terms to indicate specific objects. They do have room for subjective interpretation and would cause a bug if implemented directly into the computer program. But as a human's mind thinking as the computer, please interpret some of the general terms and specific ones.

Here is an example of a simple algorithm on how to make cereal:

1. Open the bag of cereal
2. Prepare an empty bowl that can hold more than 400mL
3. Put 30g of the cereal inside the bowl
4. Pour 200mL of milk inside the bowl

Here is another algorithm to see if you understood about the basic idea of algorithms:

1. If you truly understood algorithms, go on to read the next paragraph below.

    If else, go back to the top of the page and read the information again.

Does the above algorithm have a bug?

Although ironic to the intended meaning of this algorithm, yes it does. The statement "If you truly understood algorithms" is an ambiguous one that can be subjectively interpreted differently.

Algorithms can be found everywhere in our day to day life. Anything you do in the digital world follows an algorithm that was delicately set up by programmers. But do algorithms only exist in the digital world? The answer is no. Instructions are algorithms. Schedules are algorithms. Relationships are algorithms. To add onto that, most of our problem solving skills involve using algorithms that we set up ourselves.

2. What is an algorithm? And what are the benefits?

Believe it or not, algorithms have the possibility to save loads of time. Since words aren't fun, let's jump straight into an example.

You are handed a 600-page booklet that contains all the profiles of each student in your school. Your school contains 600 students, and each student has his/her unique profile number that ranges from 1~600. The booklet you are given has all the numbers placed in an increasing order. For example, a student with a profile number 2 has her profile in page 2.

Now your job is to find the profile of a student that has the profile number 424. What algorithm can we use to find the profile of that student?

Algorithm A:

The first algorithm might be to flip every single page from the front until we get to the number 424. This method is great, but it takes way too much time.

Algorithm B:

So maybe we can flip the pages from the back on the book, starting at page 600 and working to page 424 since we know that page 424 would be towards the second half of the book. This method is also great, and would save more time than the first method only if the number is more than 300. But it still does not seem like the 'ideal' algorithm.

The two methods shown above are surefire ways to get the end goal, but it consumes way too much time. Imagine if the booklet was more than a million pages!

Connecting to our daily life, we don't flip through every single page of the book in order to find the page we want to find. We take estimate guesses and rely on our intuition to find them.

Going back to the 600-page booklet question, if I was given the job to flip to page 424, I would first take a big estimate of where page 424 might be and flip the booklet a bit over the half. Let's say I arrived at page 390 that way. Since I know that 424 is greater than 390 but not by a big difference, I will take a smaller estimate and flip the pages. Then I arrived at page 427. Since I know 424 and 427 are very close to each other, I will start flipping the pages backwards for just a little moment until I arrive at page 424.

The procedure that I took above is way more realistic and saves a lot of time compared to Algorithms A and B. But the procedure I took involves a lot of subjective thinking that differs from each person to each other. As much as we want the computer to understand our intuitive ways of guessing, the computer can not understand these 'subjective' rules. We need to make this procedure into a set of clear rules: The algorithm.

Let's say the profile number of the student that you want to find is "N". How can we build an algorithm similar to our intuitive ways?

This is how I would have done it:

Algorithm C:

1   Open the book

2   Let "N" = the profile number of the student.

3   Turn to the middle of the book. (If the pages number is even, turn the larger page number between one of the two pages in the middle) If "Page Number" = N, finish the job.

       If else, go to step 4

4   If N is in the 1st half of the book, discard the 2nd half of the book and the page number you are on. Then go to step 3.

       If else, go to step 5

5   If N is in the 2nd half of the book, discard the 1st half of the book and the page number you are on. Then go to step 3.

       If else, the profile number of the student = N does not exist.

Just looking at the algorithm may not form an image in your mind about how this process works. Here's an example of Algorithm C using a 21-page student-profile book.

Here are the initial 21 pages of the book:

        1 2 3 4 5 6 7 8 9 10 11 12 13 14 15 16 17 18 19 20 21

We want to find the page with the student with profile number #8 in the book.

This is how Algorithm C would do it:

Step # (Relative of algorithm C)	Action taken
1	Open the book
2	N = 8
3	Open to page 11
4	Discard pages 11~21
3	Open to page 6

4	Go to step 5
5	Discard pages 1~6
3	Open to page 9
4	Discard pages 9~10
3	Open to page 8

Here is a numerical representation of what went on in that process:

(The underlined bold number indicates the number we are on)

1 2 3 4 5 6 7 8 9 10 **11** 12 13 14 15 16 17 18 19 20 21

1 2 3 4 5 **6** 7 8 9 10

7 8 **9** 10

7 **8**

We found it! Finish the job.

As seen above, Algorithm C has no room for subjective interpretation and can work with not just the 600-page or 21-page student profile book, but for every single book if we just perform a few tweaks.

Looking at Algorithm C, something may instinctively tell you that this algorithm is related to using the multiple of twos. This is very efficient compared to Algorithms A&B because when the page number of the book is doubled, Algorithm C only has to do 1 more step while Algorithms A&B have to do double the amount of steps.

When the number of pages in the book is small, there won't be a big difference in the time it takes to solve the problem. But the moment when we get stuck with a 30000-page book, it is obviously going to make a big difference. A 30000-page book would make Algorithm A&B take approximately 15000 steps each, or 30000 steps combined. Meanwhile, Algorithm C would only take $\log_2 30000$ steps.

Here is a graph that shows the efficiency of Algorithms A, B and C:

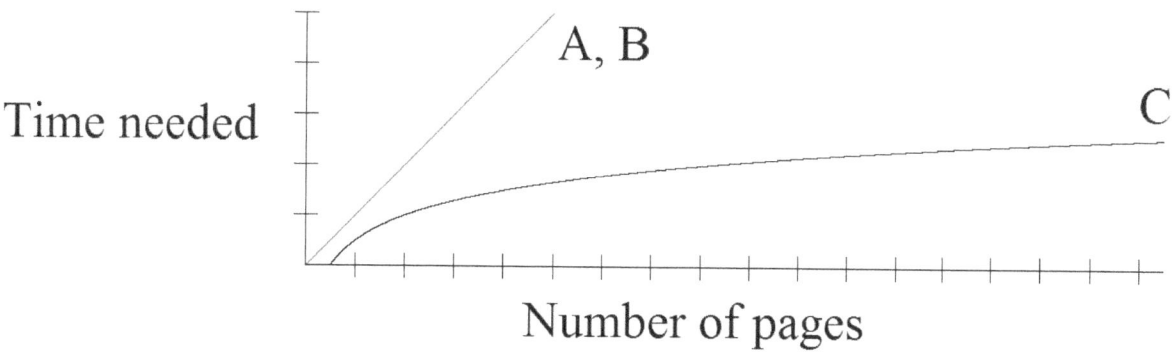

Figure 1: Algorithm speed (comparative of A, B and C, mentioned above)

You can clearly see that Algorithm C is way more useful in terms of time usage compared to Algorithm A&B. Using the example of a 30000-page book, Algorithm C would take around 2000 times faster to complete the job compared to Algorithm A&B.

So now we get the idea of what Algorithms do to make our lives much easier and how it is important to be able to manipulate a good algorithm.

3. Conclusion

There are an uncountable amount of algorithms out there in the world. Starting from basic computer science algorithms that sort numbers to algorithms that link and analyze the relationship between different entities. We've only touched on some of the basic versions of algorithms that use mathematical thinking skills involving number sorting.

I hope this short introduction to basic algorithms opened up your mind to the various capabilities of algorithms and to the significant world of computer science. There are a vast amount of resources you can access on the internet related to algorithms, and some basic programming programs such as Scratch may make you find yourself thinking very similar to those algorithm examples above.

The Rising Importance of Mathematical Statistics in Economic Fields: The Stock Market

Yongjun Lee

Chadwick International School

1. Introduction

In this paper, we will be exploring the behavior of the stock market in terms of mathematics. We can analyze this by applying mathematical techniques to an organized pool of data, a procedure known as "mathematical statistics". Some of these techniques include regression analysis (using trends to establish a broad picture of the data), distribution analysis (determining how widely spread the data is), and statistical representations (finding a single number that represents the whole data set). Today's presentation will mostly involve regression analysis as the approved method of analyzing large quantities of data.

2. Exponential Growth/Trends

When creating a general pattern or algorithm for a large set of data, a phenomenon that often arises in such cases is exponential growth. This occurs when the y-variable increases by a constant factor larger than 1 when the x-variable increases by 1. This factor is known as the base or growth factor.

One might think that this sort of graph is bizarre, but this actually occurs in an extremely high percentage of real-life situations, due to the exponential behavior of life forms. When organisms such as bacteria and humans grow, each bacterium or human has a certain number of offspring, which then also have the same number of offspring, and so on. This creates an exponential trend wherein the number of offspring for each parent becomes the growth factor. Other non-living systems, such as bank accounts and nuclear chain reactions, also show this kind of trend.

Another concept that needs explanation prior to analysis is the R2 value. This number represents the strength of a particular trend; the higher it is, the stronger the trend will be. If an R2 value is below 0.5, then its correlation is considered weak. An R2 value between 0.5 and 0.8 indicates a normal correlation, whereas anything above 0.8 is the sign of a strong correlation.

Furthermore, there are some technical nuances that make interpretation a little difficult. This is because Excel does not accept dates, and instead transforms them into serial numbers. In Excel's system, January

1, 1900 is designated as 1, January 2, 1900 is 2, January 3, 1900 is 3, and so on. Therefore, an increase of 1 in the serial number is equivalent to a day passed.

3. Analysis I: Total

In this analysis, we will use two different data sets, both from reputable stock market indexes. A stock market index is a value that attempts to measure the stock market as a whole by looking at several representative companies. Here, we will be looking at the Dow Jones and the NASDAQ, which some of you might have heard about in the news or other media.

One thing to note specifically for this analysis is that we focus on the whole graph for both stock indexes, instead of a certain time period. This allows us to look at the graph to its fullest by determining trends for the entire data set. It also prepares us for our next analysis, which will break the data into chunks; this way, we can pinpoint and later analyze particular moments in time where the correlation is very strong or becomes weak.

a) Dow Jones: 1905-Present

In the Dow, we have a century's worth of data; a full 1200 data points, one for each month. As shown in the image directly below, the trend shows an exponential graph with the exponential factor being $e^{0.0002} \approx 1.0002$ in this case. In other words, the Dow increased by a factor of approximately 1.0002 each day. The strength of this correlation can also be seen from the second picture; the exponential trend's R-squared value of 93.66 is higher than any other, including the linear and polynomial trend lines. Furthermore, we can also see that the projected trend ends far below the current index value; this can be interpreted as an acceleration of the speed with which the index increases. Another important portion of the graph to note here is the sharp drop from 2007 to 2008 during the global financial crisis, shown here by the blue arrow. Furthermore, as shown in the red frame, this seems to have highly affected the trend, driving it far down below most of the other values from the last two centuries.

b) NASDAQ: 1971-Present

Meanwhile, in the NASDAQ, a more recently published but equally trustworthy stock index, we analyze approximately 540 data points. In this trend, the exponential factor is higher ($e^{0.0003} > e^{0.0002}$), because the region from 1915 to 1970, when there is little increase in the data, has been omitted. However, the graph still shows an exponential trend, even stronger than the Dow, with a slightly higher R2 value of 0.9383. Here in the graph, we can note the sharp increase from 1994 to 2000 as the most prominent feature, as framed here in red. This increase also seems to be exponential, giving this portion more promise in terms of trends.

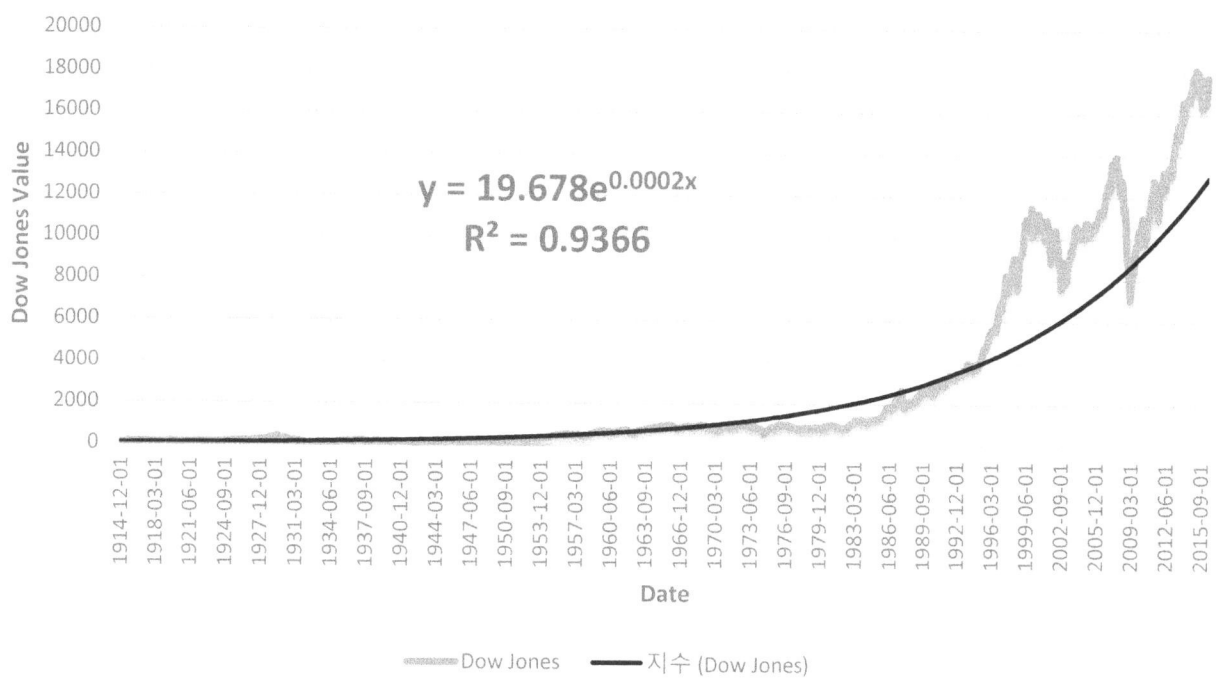

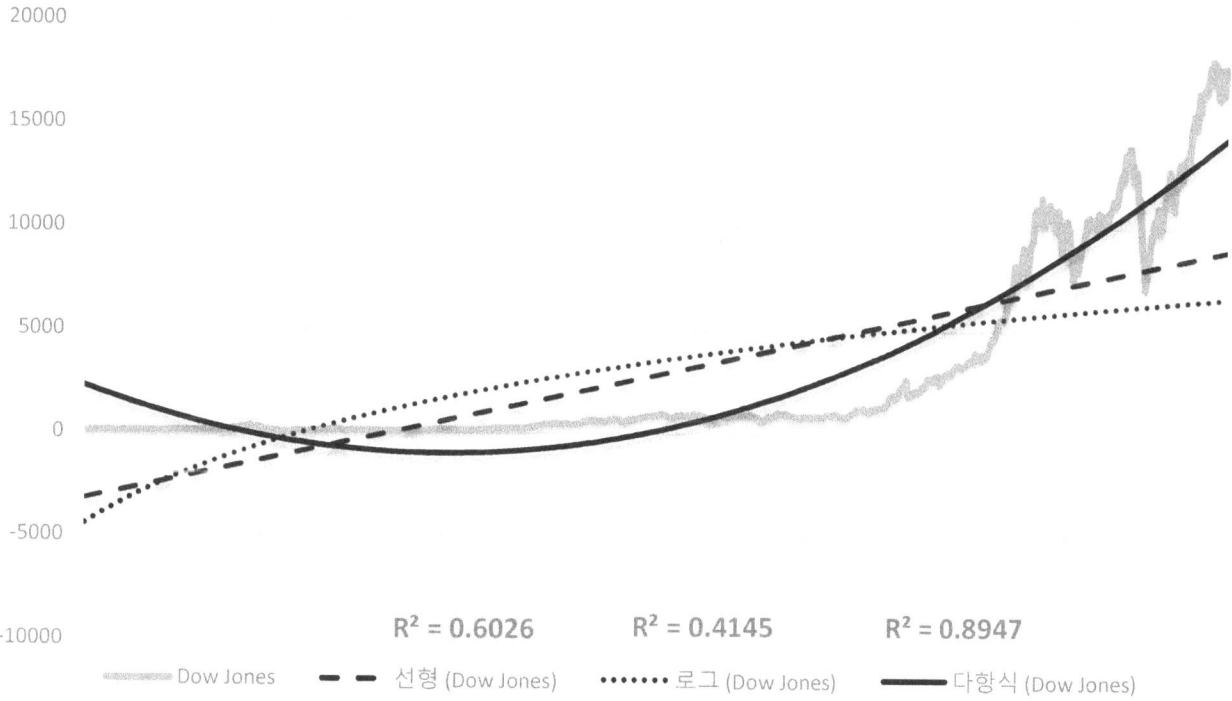

Figure 1: Dow Jones graph – analysis 1

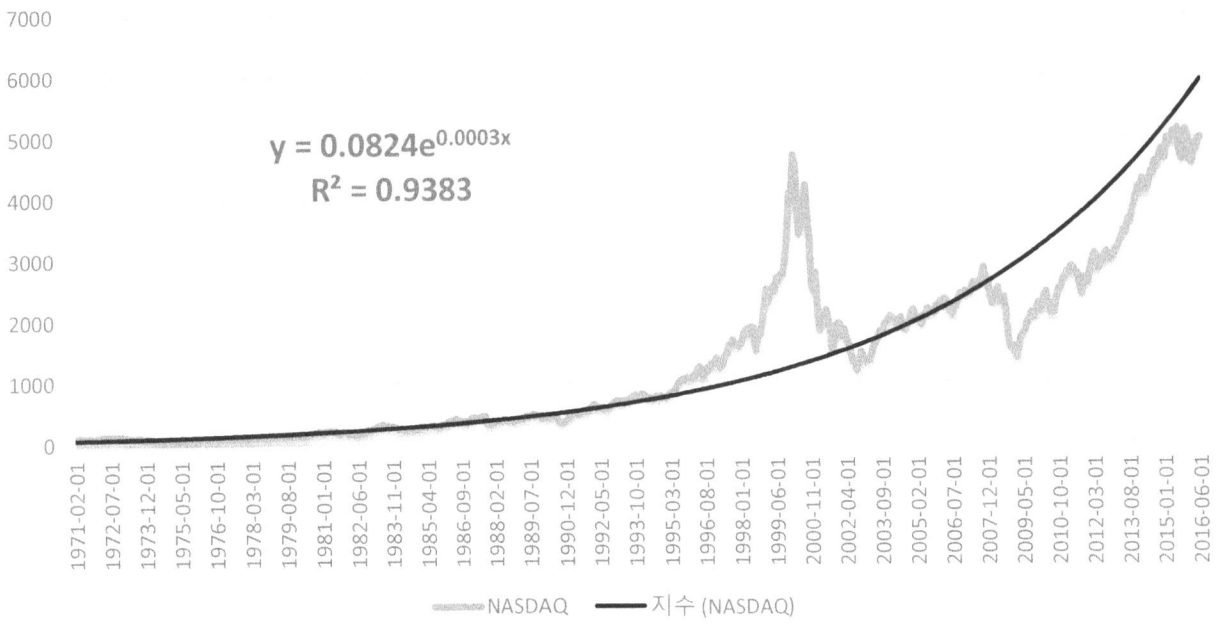

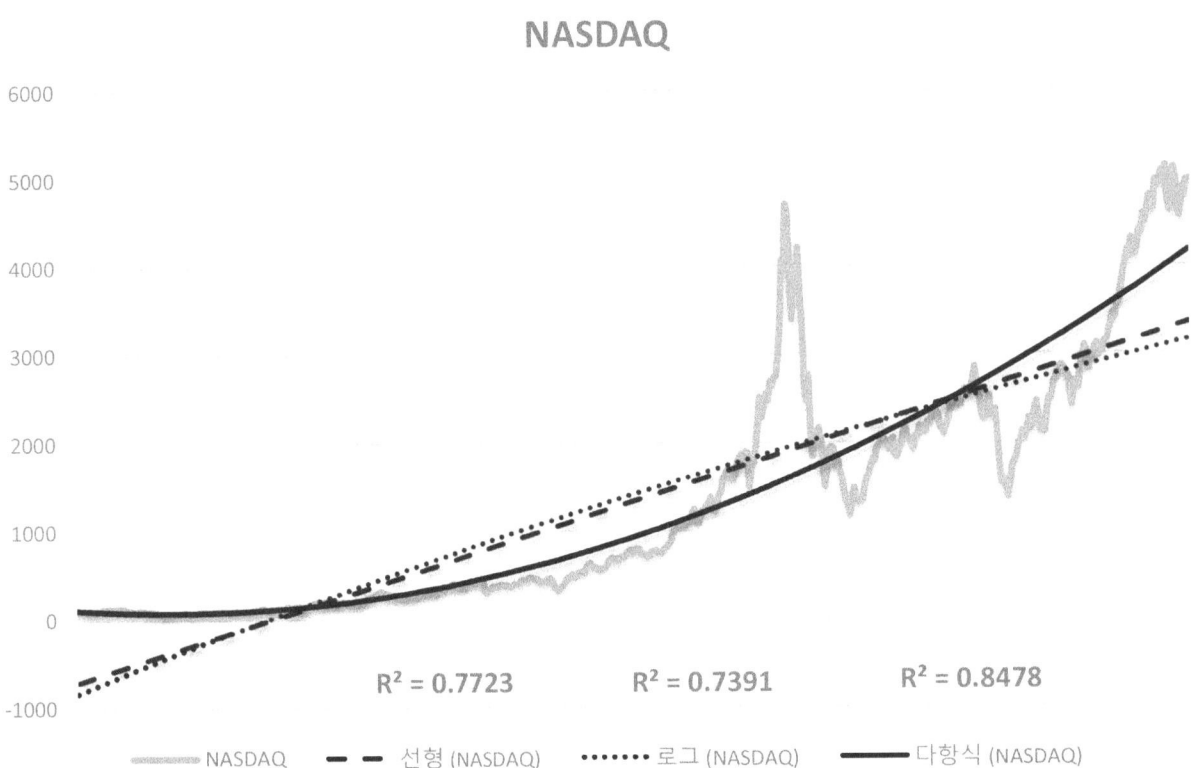

Figure 2: NASDAQ graph – analysis 1

4. Analysis II: The "Dot-Com Bubble"

In the first analysis, we looked at the graphs as a whole and identified key portions of the data that seemed statistically promising or called for further, in-depth examination. Our search unearthed two distinct key portions; the sharp increase from 1980 to 2000, and the later sharp decrease in 2007-2008. In this analysis, we will be looking at the former feature, which also coincides with an important event in global economics; the so-called "dot-com bubble."

This phenomenon began when electronics and information technology rapidly developed from the 1980s to the 2000s, providing numerous companies, namely Apple, Microsoft, Intel, and others, with very rapid growth. This resulted in huge inflation of the stock market as a whole, as new technology began to increase exponentially in value. The "bubble" eventually burst in early 2000, when several of these information tech businesses (or "dot-coms"), suddenly failed. This led to a sharp drop in the stock market overall until most of the companies involved either rebounded or declared complete bankruptcy.

a) Dow Jones: 1980-2000

In this section, we will take only the points that represent the Dow from January 1980 to March 2000, during the dot-com bubble. Here, we see that the trend clearly is exponential, with the exponential factor being similar to that of the whole graphs, but with a much higher R2 value at 0.9731, as the data is rid of outliers that were present in Analysis I.

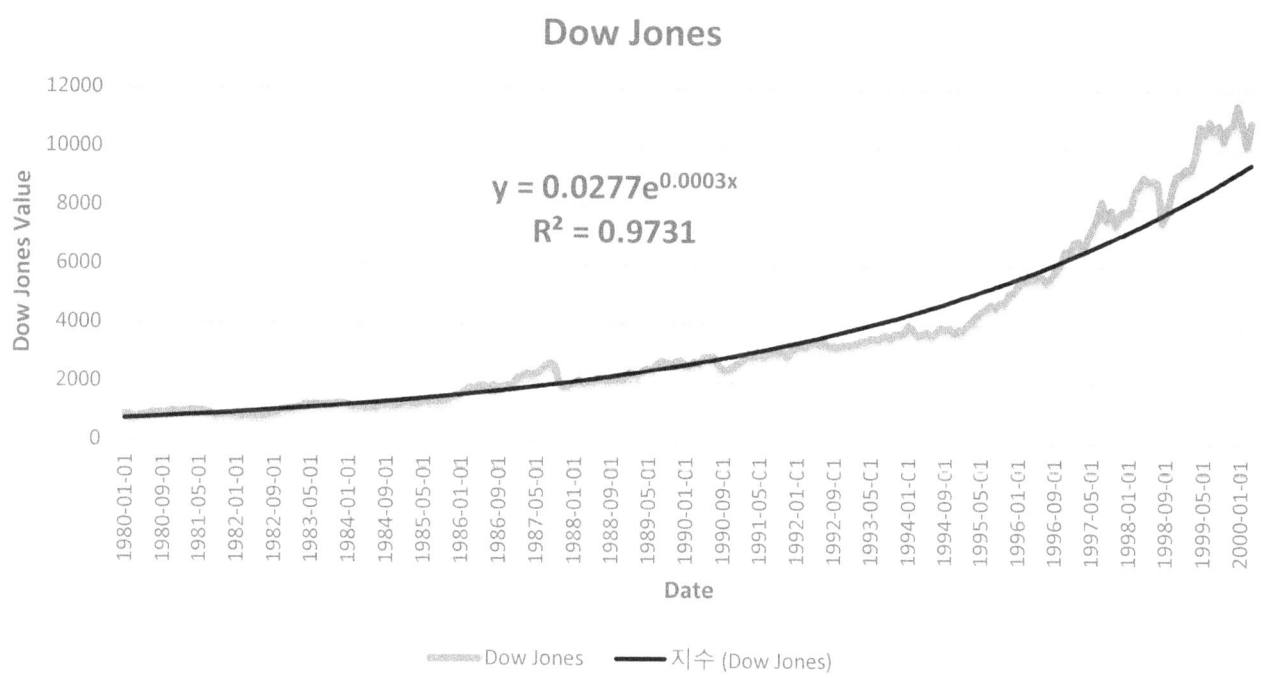

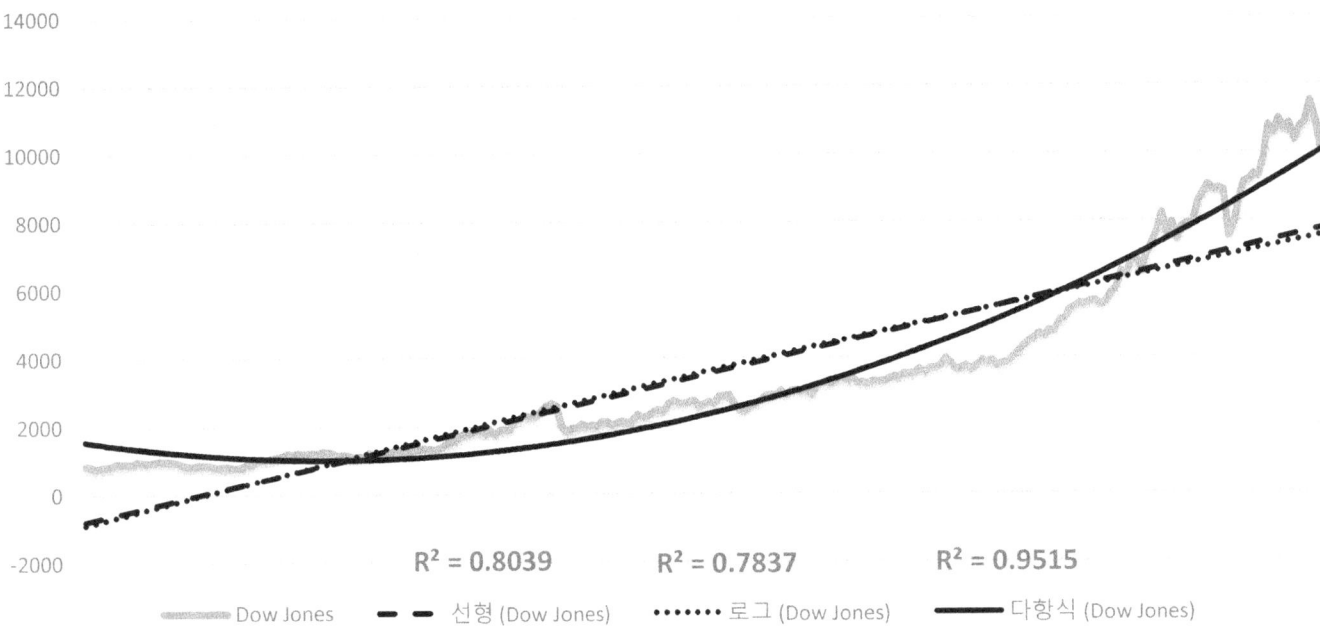

Figure 3: Dow Jones graph – analysis 2

b) NASDAQ: 1980-2000

In the NASDAQ, where the exponential increase was originally pointed out, although the visual appeal is much greater than in the Dow, the inconsistency of the increase (observe the sudden acceleration after January 1998, as shown by the green arrow) lessens the validity of the exponential trend. Nevertheless, the graph still shows a respectable R2 value and an exponential factor not far from those in Analysis I.

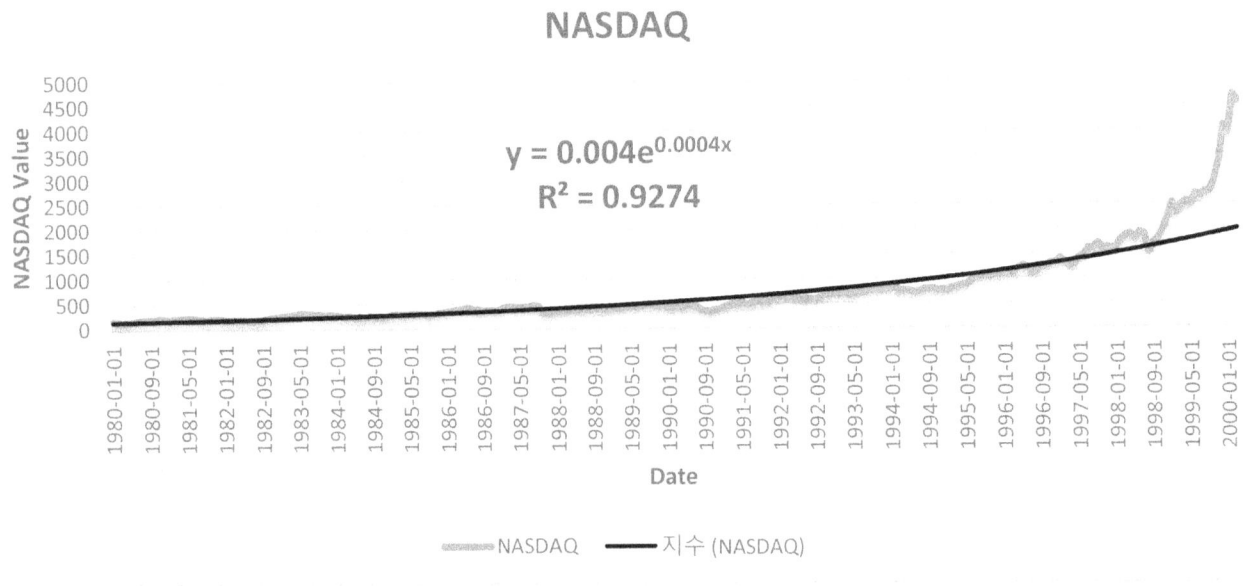

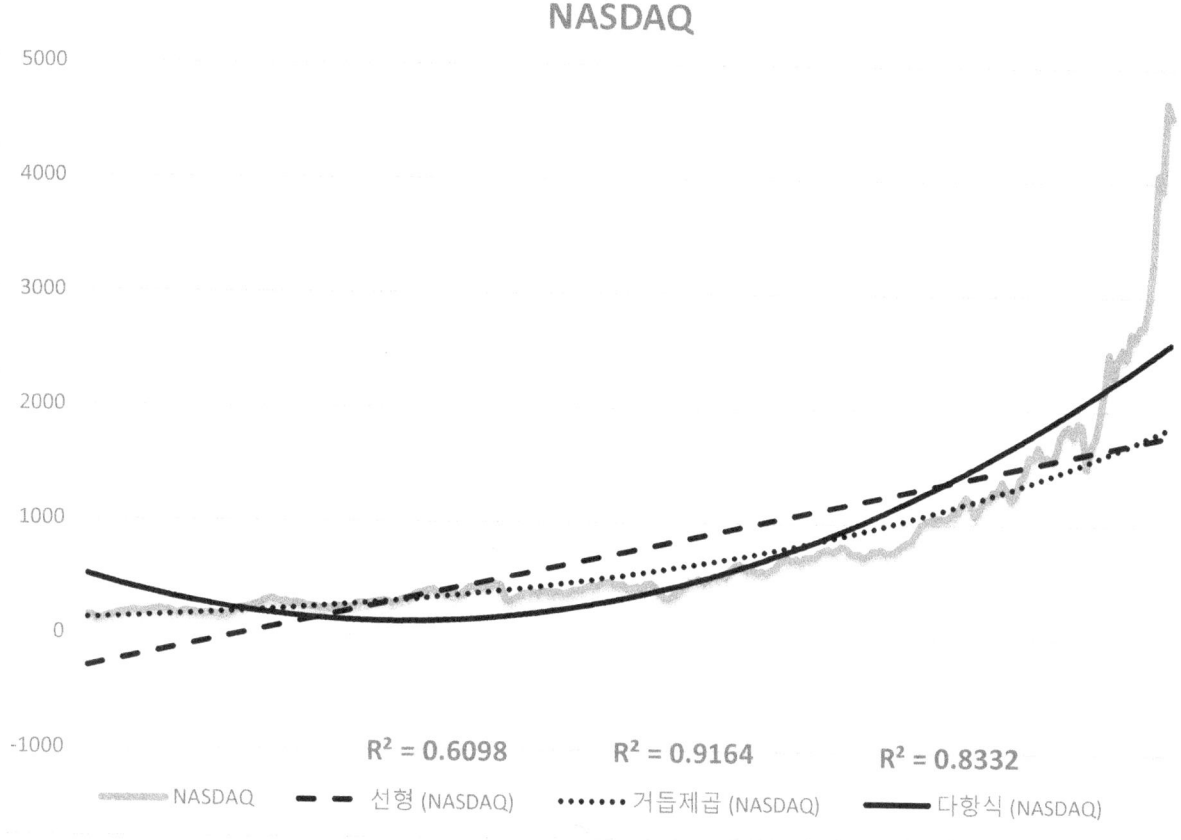

Figure 4: NASDAQ graph – analysis 2

c) S&P 500: 1980-2000

For the purposes of this analysis, I have introduced a new stock index, the Standard & Poor 500, which contains more diversified data from more companies than both the NASDAQ and the Dow Jones. Another useful point about the S&P 500 is that it publishes data sets for each day, not each month. This leads to a massive data set of more than 23,000 elements total, which is too broad for total analysis but fit for analyzing individual chunks, as here.

By taking the section from 1980 to 2000 – a full 5000 data points – and analyzing it, we can deduce a very clear exponential pattern in the data. Not only is the R2 value extremely high at 0.9677, the exponential factor is also similar to that of all the other data sets examined during this lecture; all of them are somewhere between e^0.0002 and e^0.0004. This implies that all stock indexes increase at a reasonably constant rate per day, and therefore gives rise to the idea that the stock market itself increases at such a rate. Since despite their different calculation methods, stock indexes are all representations of the market as a whole, I deduce that this theory is plausible.

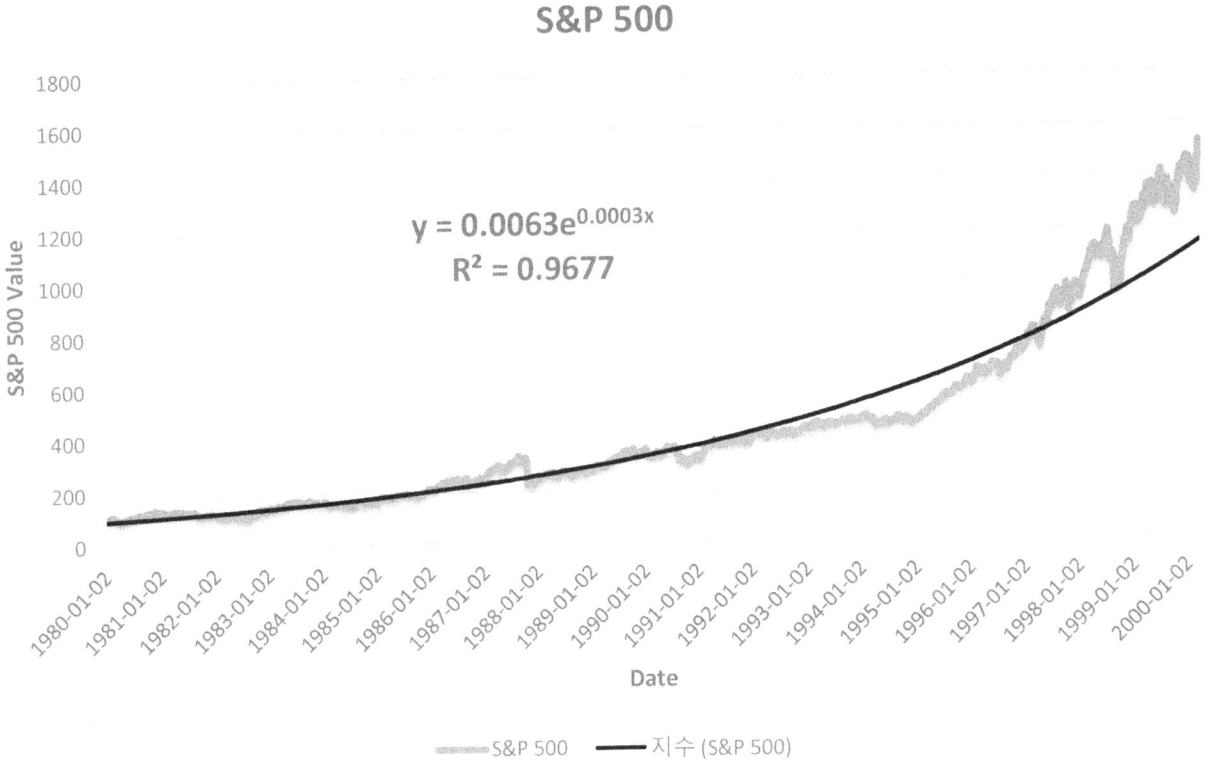

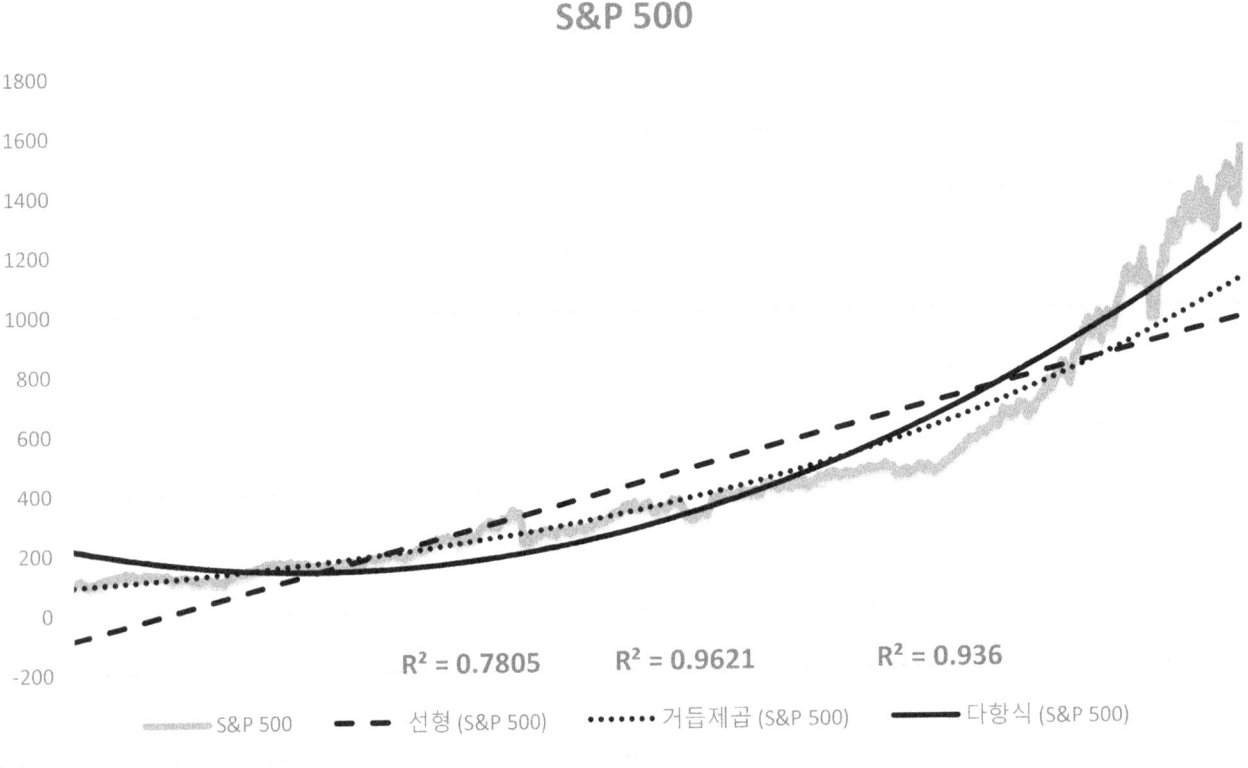

Figure 5: S&P 500: 1980-2000

d) S&P 500: 2000-2003

Now, using the stock index newly introduced in Section 4c, we can evaluate smaller chunks of it; namely, the bursting of the "dot-com bubble" from March 2000 to March 2003. This would have previously been impossible due to the sparse data; a three-year data set would only contain thirty-six data points, far too less to draw a strong conclusion. However, thanks to the day-to-day nature of Standard & Poor 500, one can look into such short-term trends with confidence as well.

Surprisingly, the decline graph shows a linear trend, the first and last we will see in this lecture. Here, according to the x-coefficient, the Standard & Poor 500 decreased around -0.6333 points per day during the bursting of the dot-com bubble. One thing to also keep in mind here is that the y-intercept 24,694 is not the starting value; according to Excel's serial number system, 2000-04-03 is 100 years, 3 months (January, February, March), and 4 days ahead of 1900-01-01, which is 1. Furthermore, January and March always have 31 days each, and 2000 is not a leap year (Multiples of 100 are not leap years in the Gregorian calendar), so therefore, the actual starting point for the x-axis is 1+365×100+31+29+31+4=36596, and the starting value for y is -0.6333×36596+24694≈1517.75. Finally, according to this graph, if the trend was to continue, we can calculate the x-intercept; x=24694÷0.6333≈38992.6. Therefore, the S&P 500 would have dropped below 0 at x = 38993, or April 22, 2008.

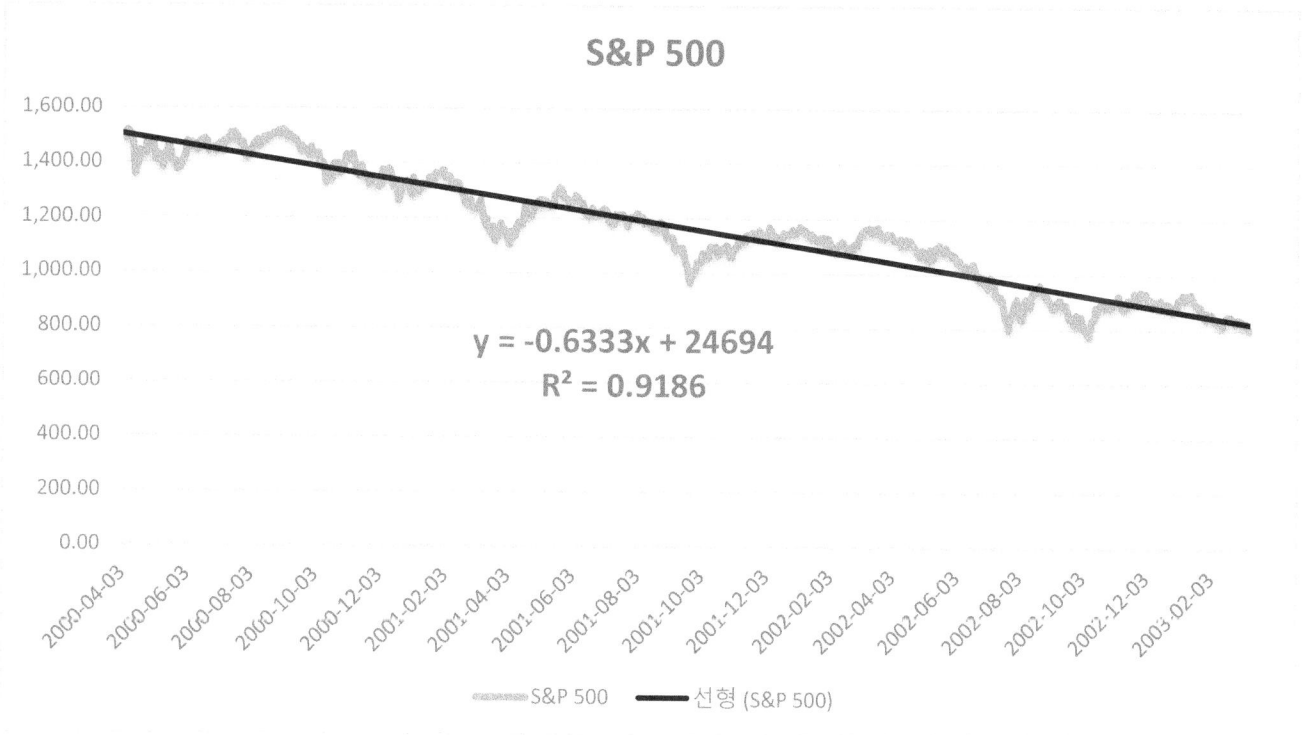

Figure 6: S&P 500: 2000-2003

5. Conclusion

Overall, in this lecture, we examined different types of correlations present in the stock market, and used them to analyze general movements of the data and point out specific regions where drastic change occurred. Later, in Analysis II, we focused on one such anomaly, which was also related to an event in economic history, the "dot-com bubble". Here, we inspected both sides of this phenomenon; the astonishing exponential increase, and the sharp linear drop that followed. From that, we were able to identify the impact that information technology had on the stock market as a whole. Furthermore, we were able to make predictions about what could have happened otherwise using the data set, and also created a small theory concerning exponential growth factors in the stock market. Finally, I hope this was a fruitful time for all of you, and thank you for reading this paper.

www.ingramcontent.com/pod-product-compliance
Lightning Source LLC
Chambersburg PA
CBHW081020240526
45471CB00017B/3449